The Fundamentals of Scientific Research

The Fundamentals of Scientific Research

An Introductory Laboratory Manual

By Marcy A. Kelly, PhD

With contributions from
Pryce L. Haddix, PhD

WILEY Blackwell

Published by John Wiley & Sons, Inc., Hoboken, New Jersey
Published simultaneously in Canada

For general information on our other products and services or for technical support, please contact our Customer Care Department within the United States at (800) 762-2974, outside the United States at (317) 572-3993 or fax (317) 572-4002.

Wiley also publishes its books in a variety of electronic formats. Some content that appears in print may not be available in electronic formats. For more information about Wiley products, visit our web site at www.wiley.com.

Library of Congress Cataloging-in-Publication data applied for

ISBN: 9781118867846

Cover image: E.coli grown on an agar plate overnight: © iStock / Getty Images Plus

Set in 11/14pt Century by SPi Global, Pondicherry, India

10 9 8 7 6 5 4 3 2

2 2016

The journey to develop this manual was linked to my personal journey from the time of its inception to its completion. There is one single person that has been my inspirational rock and source of enduring support during this endeavor. I dedicate this work to my husband, Thomas Kelly.

Contents

CONTENTS

Preface

In 2011, major stakeholders in the life sciences published a document entitled "Vision and Change in Undergraduate Education: A Call to Action" (AAAS, 2011). This document proposes a new way to teach undergraduate life sciences majors. The report suggests that undergraduate educators directly engage their students in the process of science by providing them with authentic experiences that mimic what we do as professional scientists. They postulated that biological content and the skills commonly associated with successful scientists can be introduced to students through meaningful evaluations of real biological phenomena.

The goal of the experiments in this laboratory manual is to introduce undergraduate students to the process of science through a guided introductory laboratory experience. Highlights of this manual include a semester-long experience working with one organism using experiments directed toward a single unifying goal, pre- and postlaboratory assignments and laboratory reports aimed to enhance the students' analytical, critical thinking and scientific writing skills, and experience using common laboratory equipment and techniques similar to what are used in the professional setting.

Specifically, the laboratory curriculum centers on studying *Serratia marcescens*. *S. marcescens* is a Gram-negative bacterium that is unique in that it produces a red pigment, prodigiosin, at high cell density. It has been demonstrated that prodigiosin has several interesting properties; it is antimicrobial, it has been shown to induce apoptosis in cancer cells, and it has been demonstrated that it has potential to act as an immunosuppressant (reviewed in Khanafari *et al.*, 2006). There are currently many research laboratories that are attempting to enhance the production of prodigiosin because of these unique properties. The overarching goal of the laboratory course described in this manual is to have the students learn about the organism so that they can generate and initially characterize mutants of the organism that overproduce the pigment.

The laboratory manual breaks down the laboratory course into three separate modules. For the first module, the students familiarize themselves with common laboratory equipment and techniques. For the second module, the students begin to work with and appreciate *S. marcescens* by performing growth curves and Lowry protein assays, quantifying prodigiosin and ATP production, and performing complementation studies to understand the biochemical pathway responsible for prodigiosin production. They learn how to employ Microsoft Excel to prepare and present their data in graphic format and how to use specific calculations to convert their data into meaningful numbers that can be compared across experiments. The third module requires that the students employ UV mutagenesis to generate hyperpigmented mutants of *S. marcescens* for further characterization. They use experimental data and protocols they learned during the first and second modules of the course to help them develop their own hypotheses and experimental protocols and to help them analyze their data.

For each laboratory session, students are required to answer prelaboratory questions that are designed to probe their understanding of the required prelaboratory reading materials (which includes the experimental background and protocol for that session and, in some cases, relevant primary scientific literature related to the experiments they will be performing in the laboratory). The questions also guide the students through the development of hypotheses and predictions. Following each laboratory, the students are required to answer a series of postlaboratory questions to guide them through the presentation of their data, analysis of their data, and placing of their data into the context of the primary literature. They are also asked to review their initial hypotheses and predictions to determine if their conclusions support their initial assertions. If their conclusions do not support their initial assertions, the students are asked to provide possible explanations as to why they think their conclusions did not agree with their hypotheses. The pre-laboratory and postlaboratory questions were designed to assist the students with the preparation of two formal laboratory reports after the second and third modules. The format for the reports is similar to that of primary scientific literature.

This laboratory manual seeks to introduce introductory undergraduate life sciences majors to an environment that fosters the development of scientific curiosity, creativity, and the critical thinking and communication skills required for success in the scientific disciplines. The laboratory techniques and skills that they will master through the exercises presented herein will provide the students with a strong foundation in the practice and process of science. All of these, in turn, should enable the students to persist and succeed as life sciences majors.

Acknowledgments

This work would not be possible if it was not for the students, faculty, and staff in the Department of Biology and Health Sciences–NYC. The feedback and support provided during the development of this work have been immeasurable and greatly appreciated.

About the Companion Website

This book is accompanied by a companion website:

www.wiley.com\go\kelly\fundamentals

The website includes:

- Instructor's Companion to the Lab Manual

Introduction

As life scientists, we are uniquely positioned to use our inquisitive natures to ask questions about the world around us. We have developed a systematic method to address the questions we pose. This method, the scientific method, is the major tenet behind what we do. The scientific method is as follows:

(1) Make an observation.
(2) Ask a question based upon your observation.
(3) Develop a hypothesis.
(4) Develop experiments to test your hypothesis.
(5) Collect and analyze experimental data.
(6) Confirm hypothesis or develop and test a new hypothesis.

You have probably memorized these steps at some point during your academic career, but have you ever *really* put them to the test to answer a biological question that has not yet been addressed? If not, you will gain practical experience with scientific method in this laboratory. If you have practiced the scientific method in the past, this laboratory will help you hone your skills.

Throughout the semester, you are going to be studying a single organism, *Serratia marcescens*. *S. marcescens* is a bacterium. Bacteria are considered prokaryotic organisms and have features commonly associated with prokaryotic cells. *S. marcescens* is classified as a Gram-negative bacterium because of the structure of its cell wall. Gram-negative bacteria have thin cell walls with phospholipid bilayers on both sides of the cell wall.

S. marcescens is unique among bacteria because it produces a red pigment called prodigiosin. Although the exact biological role of the pigment with respect to the organism is unknown, several laboratories throughout the world are studying the pigment and its activities. We are going to perform our own experiments in this laboratory course to understand the pigment and its biological significance. Ultimately, *your goal for this laboratory course will be to attempt to maximize the production of prodigiosin by hyperpigmented mutants that you will be creating.*

S. marcescens is ubiquitous in the environment. It preferentially grows in damp environments and is commonly seen in homes on bathroom tile grout and shower corners and at the toilet water line. When found, it appears as a pink, slimy film that is difficult to remove. In the bathroom, it primarily grows off soap and shampoo residues. Bleach-based disinfectants are typically recommended to completely remove the organism.

S. marcescens has left its mark on history several times. The Miracle of Bolsena, which occurred in the 13th century, is now attributed to *S. marcescens* contamination of the Roman Catholic host during the celebration of the Eucharist. In the Roman Catholic Church, when a priest blesses the bread and wine during the celebration of the Eucharist part of the Catholic Mass, they are believed to be transubstantiated into the body and blood of Jesus Christ. In 1263, a German priest who was celebrating Mass in Bolsena, Italy, noticed that "blood" was dripping from the host onto his linen robe and the altar linens. He attempted to clean his fingers and the altar and ended up smearing the linens further. He then took the bloodstained linens and the host to Pope Urban IV for investigation. The Pope perceived the event as a miracle that supported transubstantiation and, in honor of the miracle, created the Festival of Corpus Christi. Corpus Christi is celebrated in the Roman Catholic Church to this day.

Several microbiologists have simulated the Miracle of Bolsena on Manischewitz Passover matzos and unconsecrated wafers from both the Episcopalian and Roman Catholic churches. All samples from all laboratories involved in the study had *S. marcescens* growth on them that resembled blood (reviewed in Bennett, 1994 and Cullen, 1994).

Until the 1950s, it was believed that *S. marcescens* did not cause disease in humans—it was considered to be non-pathogenic. The US military is responsible for demonstrating that *S. marcescens* does indeed cause an atypical pneumonia in immune-compromised individuals (such as those individuals with active AIDS). The United States had an active biological weapons program from 1943 to 1969. In 1969, President Nixon declared that "... biological weapons are tactically inadequate...." He unconditionally and unilaterally renounced the testing and development of biological weapons and toxin agents—effectively shutting down the entire US program.

One reason Nixon shut down the program was from data obtained from several experiments that the US military performed on unsuspecting US populations. One such experiment, Operation Sea-Spray, was performed on September 26–27, 1950. For this experiment, the US Navy grew up large quantities of *S. marcescens*, aerosolized them, put them in balloons, and released them from barges located in the San Francisco Bay. The experiment was designed to test the effectiveness of bacterial aerosolization, the usage of balloons as biological weapons delivery systems, and the impact of wind current on dispersal distance. The US Navy selected *S. marcescens* for this test because the red pigment (prodigiosin) produced by the organism is easy to track. From this experiment, the US military concluded in a classified report that "It was noted that a successful BW [biological warfare] attack on this area can be launched from the sea, and that effective dosages can be produced over relatively large areas...."

The US military considered the experiment a success until beginning on September 29, 1950, there was a dramatic increase in the number of patients admitted to Stanford University Hospital with atypical pneumonias. Because the number of patients presenting with the atypical pneumonias continued to escalate in the following days, the US military performed a follow-up study. They demonstrated that the atypical pneumonias were all caused by the *S. marcescens* that they had sprayed over the San Francisco population. One patient, Edward J. Nevin, died from the *S. marcescens* atypical pneumonia. His family attempted to sue the US government, but the court decided that the government could not be sued because the spraying of *S. marcescens* was part of national defense planning (Cole, 2001).

S. marcescens is a very easy organism to work with. It is extremely versatile and its ability to produce prodigiosin makes it easy to detect. The laboratory course is broken down into three modules. For the first module, you will gain appreciation for and experience with common laboratory equipment and techniques that you will need for the rest of the course. For the second module, you will perform experiments to understand how *S. marcescens* grows, what it needs to grow optimally, and how and when it produces prodigiosin. You will also learn the details about prodigiosin biosynthesis and some of the biologically relevant characteristics of the molecule. For the third module, you will generate prodigiosin hyperpigmented mutants using UV light, and you will design your own experiments to initially characterize those mutants. *Your goal for the semester is to use what you learned in modules 1 and 2 to generate a hyperpigmented strain of* S. marcescens *for which you have determined the optimal conditions required for enhanced prodigiosin production.*

Using *S. marcescens* as a model organism, you will begin to build the skills to be successful undergraduate scientists. You must commit to think like a scientist by following the scientific method. Additionally, you will learn how to analyze your data and report your findings through the many writing assignments required for the course. Taken together, the work you will perform for this laboratory will provide you with a foundation for your future success as a life sciences major.

MODULE 1

Working with and Learning About Common Laboratory Techniques and Equipment

NAME ______________________________ *DATE* _____________

Exercise 1A: Using Common Laboratory Tools to Evaluate Measurements Pre-laboratory Thinking Questions

Directions

Read over the introduction and protocols for this laboratory exercise and answer the following questions to ensure that you are prepared for the session:

(1) What are the objectives for today's laboratory (provide a numbered list)?

(2) Why is it important to understand how to convert between metric units in the laboratory?

(3) Why is it important to understand how to use serological pipettes and micropipettes (what will you be using them for in the lab—the more specific, the better)?

The Fundamentals of Scientific Research: An Introductory Laboratory Manual, First Edition. Marcy A. Kelly.

Companion website: www.wiley.com\go\kelly\fundamentals

NAME ______________________________ *DATE* ____________

Exercise 1B: Using Common Laboratory Tools to Evaluate Measurements

Measurements Introduction

Based upon modifications of worksheets developed by Susan Peckham Petro, DVM.

As scientists, many of the observations we make require that we take and compare measurements. For example, in the experiment you will perform today, you will take several measurements. You will measure different volumes of water using serological pipettes and micropipettes, you will measure the mass of the water using a balance, and you will use the density of water to help you calculate percent error. As we progress throughout the semester, you will continue to realize the importance of measurement taking. We primarily use the metric system to take our measurements. The metric system, developed in France in 1791, is based on units of 10. Fractions or multiples of the standard units of length, volume, and mass have been assigned specific names. The commonly used units of the metric system are highlighted in Table 1.1. Table 1.2 shows the prefixes used to designate fractions and multiples of these commonly used units and provides examples of how they are used.

Table 1.1 Commonly used units of the metric system.

Measurement	Unit
Length	Meter (m)
Volume	Liter (l)
Mass	Gram (g)
Molar	Concentration (M)

Source: Susan Peckham Petro, DVM.

Table 1.2 Prefixes used to designate fractions and multiples of these commonly used units.

Fraction or multiple	Prefix	Symbol	Common usage of prefix
1×10^{6}—one million	Mega	M	.
1×10^{3}—one thousand	Kilo	K	.
1×10^{-1}—one tenth	Deci	D	Decade has 10 years
1×10^{-2}—one hundredth	Centi	c	Century has 100 years
1×10^{-3}—one thousandth	Milli	m	Millennium has 1000 years
1×10^{-6}—one millionth	Micro	µ	.
1×10^{-9}—one billionth	Nano	n	.

Source: Susan Peckham Petro, DVM.

Often, upon taking a measurement, you will be required to convert that measurement to units that are different than the ones you used to initially take the measurement.

To Convert Smaller Units to Larger Units: Divide by the appropriate factor of 10 because there are fewer of the larger units.

Example 1: According to Table 1.2, a millimeter (milli = one thousandth) is 10 times smaller than a centimeter (centi = one hundredth). To change 1 millimeter (mm) to centimeters (cm), you must divide 1 by 10. So 1 mm is equivalent to 0.1 cm.

Example 2: According to Table 1.2, a nanogram (nano = billionth) is 1000 times smaller than a microgram (micro = millionth); therefore, to change 1 nanogram (ng) to micrograms (ug), you must divide 1 by 1000. So 1 ng is equivalent to 0.001 ug.

To Convert Larger Units to Smaller Units: Multiply by the appropriate factor of 10 because there will be more of the smaller units.

Example 1: According to Table 1.2, a decimeter (deci = one tenth) is 100,000 times bigger than a micrometer (micro = one millionth); therefore, to convert 1 decimeter (dm) to micrometers (um), you must multiply 1 by 100,000. So 1 dm is equivalent to 100,000 um.

Example 2: According to Table 1.2, a meter is 1000 times larger than a millimeter (milli = thousandth); therefore, to convert 54 meters (m) to millimeters (mm), you must multiply 54 by 1000. So 54 m is equivalent to 54,000 mm.

The Shortcut for Metric Conversions: The shortcut is simple—move the decimal points! When you realize that the units are 1000-fold apart, you can move the decimal over three places (for the three 0's) and change the units. Make sure that you are moving the decimal point the correct way when you do this:

If you convert from a larger unit to a smaller unit, move the decimal to the right. For example, if you want to convert 2.001 milliliters (ml) to microliters (ul), you must first recognize that there is a 1000-fold (1×10^3) difference between the two units and then move the decimal place three positions to the right (one position for each 0 in 1000) as indicated in the equation below:

Equation 1.1

Metric conversion from larger units to smaller units.

$$2.001 \text{ ml} = 2001 \text{ ul}$$

If you convert from a smaller unit (e.g., ng) to a larger unit (e.g., gram), move the decimal to the left. For example, if you want to convert 51,000 nanograms (ng) to grams (g), you must first recognize that there is a one billion-fold (1×10^9) difference between the two units and then move the decimal place nine positions to the left (one position for each 0 in one billion; you will have to add 0's to the left side of the number) as indicated in the equation below:

Equation 1.2

Metric conversion from smaller units to larger units.

$$000051{,}000.\ \text{ng} = 0.000051\ \text{g}$$

Scientific Notation: As you can see from some of the examples above, some of the numbers have many digits. Scientific notation is a method used by scientists to simplify those numbers. Let's convert the numbers provided in the metric conversion shortcut examples above to scientific notation.

To convert a whole number to scientific notation, place a decimal point to the right of the first digit in the number. To determine the exponent, count the number of digits to the left of the decimal point you just added. For example:

Equation 1.3

Conversion of a whole number into scientific notation.

$$2001\ \text{ul} = 2.001 \times 10^3\ \text{ul}$$

To convert a decimal number to scientific notation, place a new decimal point to the right of the first whole digit in the decimal. To determine the exponent, count the number of digits between the old and new decimal point. The exponent should be a negative number. For example:

Equation 1.4

Conversion of a decimal into scientific notation.

$$0.000051\,\text{g} = 5.1 \times 10^{-5}\,\text{g}$$

Significant Figures: Significant figures are the number of digits required to express the result of a measurement so that only the last digit in the number is in question. This means that when you are recording measurements, you should include all of the digits you are sure of *plus* a rounded estimate of the next smaller digit to the nearest tenth. In practice for the exercises in this laboratory manual, use, at most, three significant figures when reporting measurements. Below, please find some rules for determining the number of significant figures in a measurement:

(1) The number of significant figures does not change when the decimal point is moved.

Example: 625.2 m written as 0.6252 km—both have four significant figures

(2) Zeros between two significant digits are always significant.

Examples: 5.00004 has six significant figures, 94203 has five significant figures, and 650.007 has six significant figures

(3) Trailing zeros to the right of the decimal point are significant in every measurement.

Examples: 5.00 has three significant figures, 27.0 has three significant figures, and 30000.0 has six significant figures (You need to consider rules 2 and 3 for this last example as well, do you know why?).

(4) Leading zeros are not significant in any measurement.

Examples: 0.00007 has one significant figure, 0.708 has three significant figures, and 0.07808 has four significant figures (You need to consider rules 2 and 3 for this last example as well, do you know why?).

(5) Trailing zeros appearing to the left of the decimal point may not be significant.

Examples: 500 has at least one significant figure, 54640 has at least four significant figures, and 87,090,000,000 has at least four significant figures.

(6) Any zeros that disappear when you convert a measurement to scientific notation are not significant.

Examples: 4000 converted to scientific notation (4×10^3) has at least one significant figure, 0.00064 converted to scientific notation (6.4×10^{-4}) has two significant figures, 260.00400 converted to scientific notation (2.6000400×10^2) has eight significant figures, and 0.008090 converted to scientific notation (8.090×10^{-3}) has four significant figures.

Let's Put This to Practice!

Work independently to perform the following metric conversions. Your answers, where appropriate, should include three significant figures and be written out in scientific notation. After you complete the following problems, work with your laboratory group to come to a consensus for each answer. Be sure everyone in your group understands how each answer was determined. Finally, discuss your results with the class.

Convert 100 um to mm.

Convert 251 mg to g.

How many ug are in 354 kg?

How many seconds are in 2.567 ns?

Which is smaller—ul or ml?

Which concentration is higher—M or mM?

Which is larger—nm or mm?

Write 25 ug in g.

Convert 0.075 ml to ul.

Convert 659 nm to um.

Working with Serological Pipettes and Micropipettes

A vast majority of the exercises that you will work on in this laboratory require that you are proficient in using different tools to measure and dispense specific volumes of liquids. These tools include disposable serological pipettes and micropipettes.

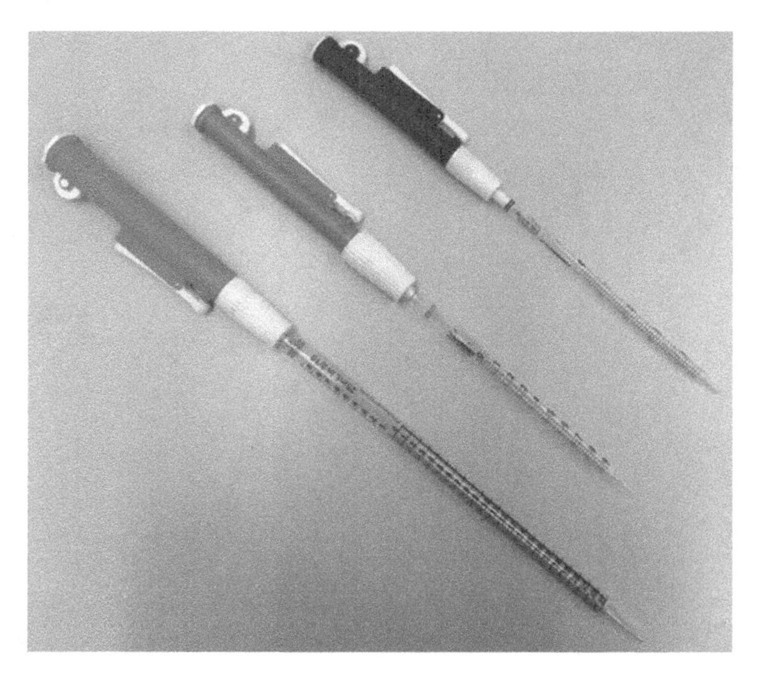

Figure 1.1 Serological pipettes and pumps for total volumes of 25 ml (red pump), 10 ml (green pump), and 5 ml (blue pump). (*See insert for color representation of the figure.*)

There are four types of serological pipettes that you will be using in the laboratory based upon the total volume that they can draw up: 2, 5, 10, and 25 ml. Each of the four serological pipettes is calibrated so that you might be able to ensure that you draw up the appropriate volume of fluid as per the protocol you are working with. In order to draw up fluid, a pipette pump is attached to the top of the appropriate pipette to provide suction. Your laboratory instructor will demonstrate how to attach the pipette pumps and use the serological pipettes. Figure 1.1 provides examples of serological pipettes with attached pipette pumps.

To measure smaller volumes of fluid (<1 ml), micropipettes are used. Most micropipettes are considered adjustable micropipettes—meaning that you can manually adjust the volumes you wish to draw up on the micropipette itself. You will be working with three

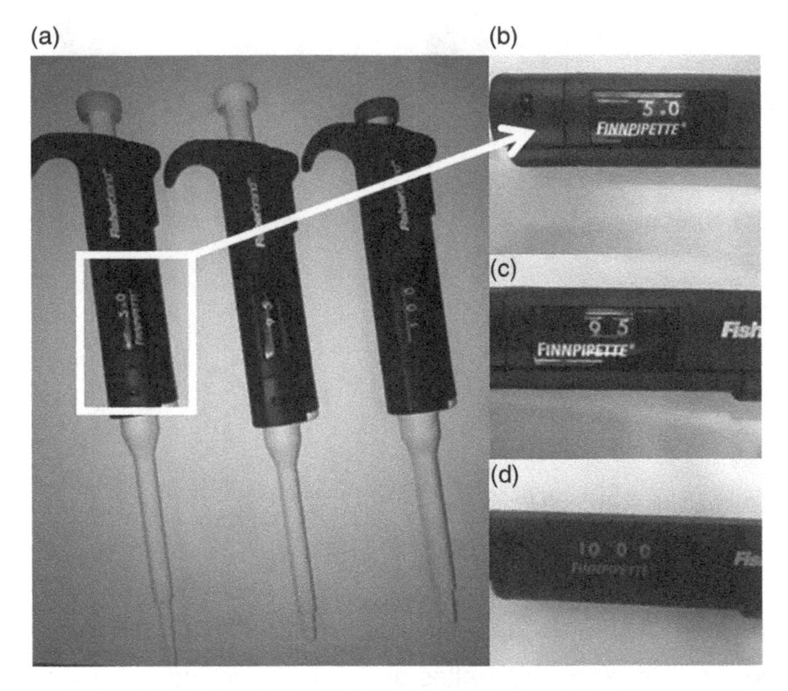

Figure 1.2 (a) Typical 10, 100, and 1000 ul total volume micropipettes, respectively. (b) 10 ul micropipette volume indicator. The volume indicator on a 10 ul micropipette is read from left to right. Digits to the left of the decimal point indicate uls, and digits to the right of the decimal point indicate tenths of uls. (c) 100 ul micropipette volume indicator. The volume indicator on a 100 ul micropipette is also read from left to right. Digits indicate uls up to 100 ul. (d) 1000 ul micropipette volume indicator. The volume indicator on a 1000 ul micropipette is read from left to right. Digits indicate uls up to 1000 ul. (*See insert for color representation of the figure*.)

different micropipettes in this laboratory based upon the volumes they can draw up: 1000 ul to 100 ul, 100 ul to 10 ul, and 10 ul to 0.5 ul. In order to draw up fluid using the micropipettes, you will need to attach the appropriate disposable plastic pipette tips to the end of the micropipettes. Your laboratory instructor will demonstrate how to attach the tips and use the micropipettes. Figure 1.2 provides an example of micropipettes.

Let's Put This to Practice!

Serological Pipettes: Work in groups of four for this exercise.

(1) Obtain a clean, dry 250 ml beaker.

(2) Place the beaker on the electronic balance provided to your group and use the tare control on the balance to adjust the reading to 0. You need to do this to compensate for the mass of the empty beaker so that its mass is not included in your readings. Your instructor will show you how to tare if you have difficulty.

(3) Obtain a flask and add approximately 60 ml of tap water to it.

(4) Using the serological pipettes, dispense the water into the beaker on the balance following Table 1.3. When you do this, be sure to use a brand new serological pipette for each volume dispensed. In addition, please make sure that the beaker remains on the balance and that the balance is on when you add the water.

The density of water (0.998 g/ml at 23°C) can be used to check the accuracy of your ability to dispense the water into your beaker using

Table 1.3 Volumes to dispense to practice using serological pipettes.

Pipette type (ml)	Volume to dispense (ml)
25	19
10	8
10	6
10	10
5	2
5	5
Total expected volume	*50*

the serological pipettes. Answer the following questions to determine how accurate your pipetting was:

(1) What was the mass of the water in the beaker upon completion of all of the pipetting? _____

(2) Using the density of water what is the actual volume of water in the beaker?

(Hint: volume of water = mass/density) _____

(3) Finally, to determine how accurate your pipetting was, you can determine your % error. The equation for % error is as follows:

$$\% \text{ error} = (\text{dispensed value} - \text{expected value}) / \text{expected value} \times 100$$

What is your percent error (keep in mind that you may end up with a negative number if your actual dispensed volume was lower than the expected volume—Why is this?)? _____

Micropipettes: Work in groups of four for this exercise. Micropipettes are very expensive and delicate pieces of laboratory equipment. *Never* exceed the upper and lower volume limits of the micropipettes!

(1) Hold the micropipette in one hand. With the other hand, turn the volume adjustment knob to the desired setting (Figure 1.2).

(2) Attach a new disposable tip to the pipette shaft. Be sure the tip is properly attached and has a good seal.

(3) Press the plunger to the first stop (Figure 1.3b) where you feel a slight resistance. This represents the volume displayed on the digital indicator (Figure 1.2).

(4) Holding the micropipette vertically, immerse the tip a few ml into the sample you wish to suck up while holding the plunger at the first stop.

(5) Allow the plunger to slowly return to the UP position (Figure 1.3a). Remember do not let it "snap" to the UP position. Then carefully withdraw the tip from the sample making sure there are no air bubbles.

(6) To dispense the liquid to a new tube, gently touch the tip to the side of the receiving vessel. Press the plunger past the first stop to the second stop (Figure 1.3c). With the plunger fully pressed, withdraw the tip carefully, wiping residual drops against the vessel wall.

(7) Allow the plunger to slowly return to the UP position.

(8) Discard the tip by depressing the tip ejector button.

Practice Pipetting Small Volumes

(1) Pipette and then dispense 10.0 ul and 5.0 ul samples of water onto a piece of Parafilm using the 0.5–10 ul micropipette. Be sure that both water samples end up in the same "bubble" on the Parafilm.

(2) Use the 10–100 ul micropipette to draw up the 15.0 ul on the Parafilm. Keep practicing this until you can do this without leaving any liquid behind on the Parafilm and until no air is introduced into the tip.

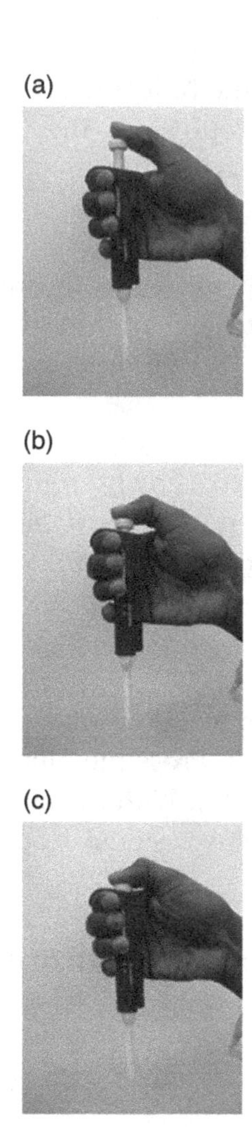

Figure 1.3 (a) Up position of the plunger button on a micropipette. This is the starting position for proper pipetting. (b) The plunger button is depressed to the first stop position to initiate pipetting. While holding the plunger button in this position, insert the micropipette tip into the sample and then slowly release the plunger button to the up position. The desired volume of sample will be drawn up into the tip. (c) When the plunger button is depressed beyond the first stop position to the second stop position, the liquid in the pipettor tip will be completely expelled. (*See insert for color representation of the figure.*)

(3) Pipette and then dispense 6.2 and 8.8ul samples of water onto the Parafilm, but, this time, do not combine both water samples into a single "bubble."

(4) Use the 10–100ul micropipette to draw up the two bubbles by depressing the plunger on the micropipette to the first stop. While releasing the plunger very slowly, draw up both the 6.2ul volume and the 8.8ul volume within one release (stroke) of the plunger. Keep practicing until you can do this without leaving any liquid behind on the wax paper. If air is allowed to be introduced into the micropipette after the 8.8ul volume, you will not be able to draw up all the remaining liquid.

NAME ______________________________ *DATE* ____________

Exercise 1C: Using Common Laboratory Tools to Evaluate Measurements Post-laboratory Thinking Questions

Based upon modifications of worksheets developed by Susan Peckham Petro, DVM.

Directions

Answer the following questions upon completion of the laboratory exercises:

Complete Table 1.4 by converting the metric linear measurements to their English equivalents (see below). Both should be expressed with no more than three significant figures, if appropriate, and in scientific notation.

Conversion between the English and Metric Systems

1 in. equals 2.5 cm.

1 mile equals 1.6 km.

1 m equals 39 in.

Table 1.4 Conversion of metric to English units of measurements.

Metric unit	Definition in meters	English equivalent
20.0 kilometers (km)		miles
0.5 centimeters (cm)		inches
1.0 decimeter (dm)		inches
1.0 millimeter (mm)		inches
1.0 micrometer (um)		inches
1.0 nanometer (nm)		inches

NAME ______________________________ *DATE* ____________

Exercise 2A: Using Microscopy to Evaluate Cell Size and Complexity Pre-laboratory Thinking Questions

Directions

Read over the introduction and protocols for this laboratory exercise and answer the following questions to ensure that you are prepared for the session:

(1) What are the objectives for today's laboratory (provide a numbered list)?

(2) In your own words, describe the steps you would take to focus on a specimen that requires the 40× objective to be seen with the microscope. Be sure to use the correct terminology when presenting your points.

(3) What is your hypothesis with respect to the type of cells you will observe in the laboratory compared to cell size? Support your hypothesis with facts.

(4) In your own words, describe why it is important that you use sterile technique when working with bacteria in the laboratory.

The Fundamentals of Scientific Research: An Introductory Laboratory Manual, First Edition. Marcy A. Kelly.

Companion website: www.wiley.com\go\kelly\fundamentals

NAME ______________________________ *DATE* ____________

Exercise 2B: Using Microscopy to Evaluate Cell Size and Complexity

Microscopy Introduction

Why Microscopes: The naked eye is unable to detect anything smaller than 0.1 mm in diameter. Therefore, most cells, tissues, and small organisms (such as bacteria) cannot be observed directly. The light microscope is required to view most plant and animal cells, the eukaryotic cell nucleus and mitochondrion, and most bacterial cells. If you wish to view organisms that are smaller than 1.0 um in diameter, you must use an electron microscope.

The light microscope is perhaps the single most important instrument used in cell biology. It can be used under bright field conditions to study the organization of cells in fixed and stained sections of tissues. With phase contrast optics, it is possible to monitor the movements of living cells and to observe changes in their subcellular organization. The light microscope may also be used to monitor certain experimental techniques such as cell fractionation and biochemical characterization of cellular components.

Magnification versus Resolution: Good microscopy depends upon the capacity of the microscope to resolve objects, not magnify them. Magnification is the ability to enlarge an object. Resolution is the ability to clearly distinguish two points that are close together within an object. Therefore, a microscope with high resolving power enables individuals to view magnified details more clearly, whereas microscopes with low resolving power produce blurry images. The image depicted in Figure 2.1 illustrates the relationship between magnification and resolution. It is an image of bacteria magnified

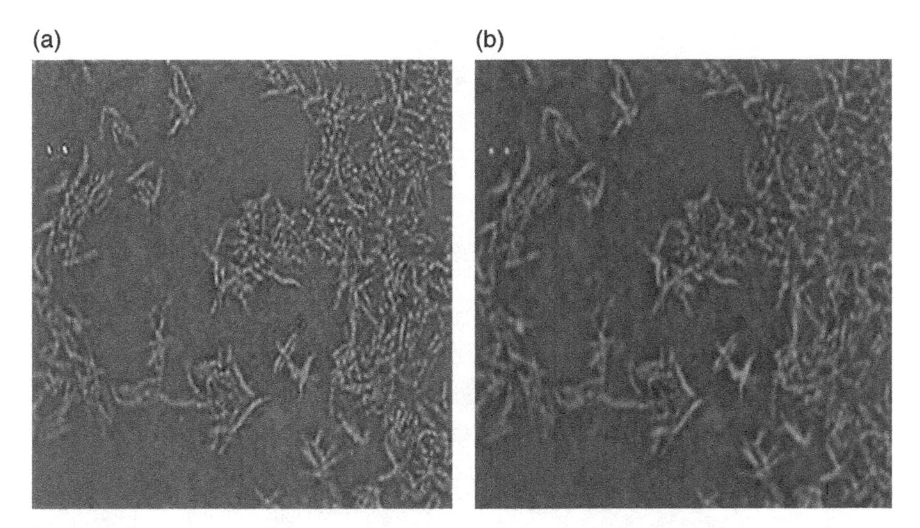

Figure 2.1 Example of an image of bacteria magnified 1000× with a light microscope with high resolving power (a) and low resolving power (b).

1000× with a light microscope. The panel to the left (Figure 2.1a) is an image magnified with high resolution. The panel to the right (Figure 2.1b) is the same magnified bacteria with low resolution. Therefore, without resolving power, magnification is pointless.

Microscopy and Contrast: A microscopic specimen cannot be viewed using a microscope unless there is significant contrast between the specimen and the background. Contrast is due to the differences in light absorption between different parts of a specimen. A vast majority of biological specimens are translucent and therefore require staining to be viewed under a microscope. Staining is frequently used to enable visualization of an organism or components of an organism under the microscope. If bacteria are not stained, they cannot be detected using microscopy.

Components of a Light Microscope: You will use the compound light microscope throughout your academic careers as undergraduate biology students. The microscope has several important features

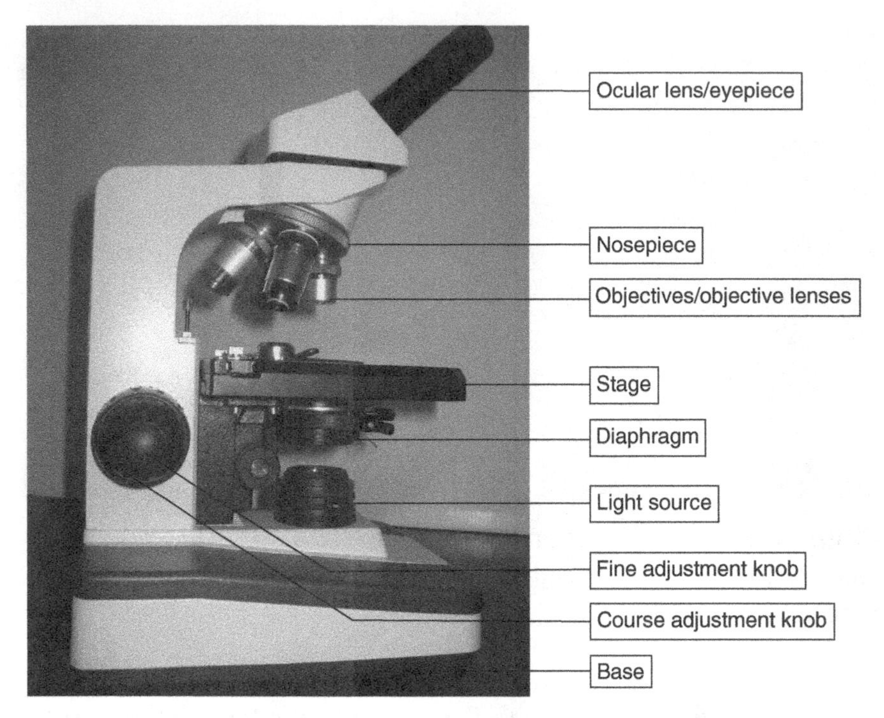

Figure 2.2 Components of the compound light microscope. (*See insert for color representation of the figure*.)

that you are responsible for becoming familiar with. Figure 2.2 and the text that follows highlight those features:

(1) Oculars lens or eyepieces—This is the lens you look through when using a microscope. Microscopes with a single ocular lens are considered monocular, whereas microscopes with two ocular lenses are called binocular. Most ocular lenses have magnifications of 10×. Ocular lenses can also be equipped with pointers and measuring scales called ocular micrometers.

(2) Nosepiece—Component of the microscope that aids in the positioning of the objective lenses.

(3) Objectives or objective lenses—Most light microscopes have sets of three or four objective lenses. The objective lenses project light from the specimen through the body of the microscope to the oculars so that you might view the specimen. Each objective lens has a different magnification, and the magnifications are usually stamped on the objectives. These magnifications include 4× (scanning magnification), 10×, 40×, and 100× (oil immersion). In order to determine the total magnification that is being used to view a specimen, remember that the ocular lens has a magnification of 10×. Therefore, you need to multiply the magnification of the ocular lens (10×) by the magnification of the objective lens to determine the total magnification that you are using.

(4) Stage—This is the horizontal surface upon which the slide containing your specimen is placed. The stage is equipped with stage clips to hold the slide in place as you are viewing it. There is a knob (typically positioned below and behind the stage) that can be used to move the stage vertically.

(5) Diaphragm—The diaphragm is a component of the substage condenser system. This system focuses light on the specimen. The diaphragm is an adjustable light barrier in the condenser. There are two types of diaphragms—iris or annular diaphragms. An iris diaphragm opens and closes the condenser with a smooth movement using a lever. An annular diaphragm is composed of a plate under the stage that, upon rotation, places circles of varying diameters in the light path. This controls the amount of light that illuminates the specimen.

(6) Light source—This component is responsible for shining light on the specimen. It is located below the stage and is controlled by an on/off switch. Many microscopes are also equipped with the ability to control the intensity of light from the light source using a lever.

(7) Coarse adjustment knob—Once you have located your specimen on the slide, this knob is used to roughly focus on the specimen. This knob should only be used when using the scanning (4×) or 10× objectives.

(8) Fine adjustment knob—The knob is used for fine focusing on the specimen. Once you have located the specimen using the coarse adjustment knob, this knob can be used. Many microscopes are parfocal. This means that once you fine focus the specimen at a lower magnification and switch to the next higher magnification, there will be no need to refocus on the specimen using the coarse adjustment knob. You might have to slightly adjust the fine focus but, not by much.

(9) Base—Cast metal component of the microscope upon which the entire scope is mounted.

Let's Put This to Practice!
Several slides will be available to you in the laboratory to use in order to practice using the light microscope. Each student should view the slides at the maximum total magnification listed (see below for the procedures explaining focusing to view each of the slides):

Prepared slide of mixed bacteria—400×

Prepared slide of *Paramecium*—400×

Prepared slide of *Saccharomyces cerevisiae*—400×

Prepared slide of an onion root tip—100×

Prepared slide of human skin cells—100×

Each of the slides that you will look at contains cells from organisms that belong to the different groups in the Linnaean classification system of organisms. This classification system breaks down organisms into three distinct domains based upon genetic, structural, and/or functional similarities. The three domains are Archaea, Bacteria, and Eukarya.

The cell theory, published by Theodor Schwann and Matthias Jakob Schleiden, states that

> "The cell is the basic unit of life and that all living organisms are composed of one or more cells or the products of cells."

As such, organisms in each of the three domains are composed of cells, but the cells differ greatly in their structure and function.

Biologists recognize two different types of cells—prokaryotic cells and eukaryotic cells. The members of domains Archaea and Bacteria contain prokaryotic cells, and the members of domain Eukarya contain eukaryotic cells. Prokaryotic cells lack nuclei, chromosomal proteins, and membrane-bound organelles. Eukaryotic cells contain everything that the prokaryotic cells lack. Of the slides you will view, the slide containing the mixed bacteria contains cells belonging to domain Bacteria. Therefore, they are prokaryotic cells. The other four slides that you will view contain eukaryotic cells and represent the different kingdoms in domain Eukarya. These kingdoms, listed in increasing complexity, include Protista, Fungi, Plantae, and Animalia. The slides, their kingdom classification, and some unifying features of organisms belonging to each kingdoms are listed Table 2.1.

Procedure: Steps Used When Viewing a Slide

(1) Obtain a microscope, plug it in, and turn it on.

(2) Clean the ocular and objective lenses with lens paper.

EXERCISE 2B

Table 2.1 Characteristics of members of the kingdoms represented by the prepared slides viewed in this exercise.

Slide	Kingdom	Unifying features
Prepared slide of mixed bacteria	Eubacteria	Single-celled prokaryotes with cell walls. Exist as different shapes such as rods or spheres and can form patterns including chains and clusters. Motile bacteria use either cilia or flagella
Prepared slide of *Paramecium*	Protista	Single-celled or multicellular aquatic organisms that can either be nonmotile or motile. Motile protists use either cilia, flagella, or pseudopods for movement
Prepared slide of *Saccharomyces cerevisiae*	Fungi	Single-celled yeast or multicellular mold. All fungi are absorptive heterotrophs with cell walls composed of chitin. They are commonly found on decaying materials
Prepared slide of an onion root tip	Plantae	Multicellular organisms that convert light energy into sugar by a process called photosynthesis. These organisms all contain cell walls composed of cellulose
Prepared slide of human skin cells	Animalia	Multicellular, motile organisms that support an ingestive heterotrophic mode of nutrition. All animal cells lack cell walls

(3) Place the slide that you are going to view on the stage and clip it in place with the stage clips.

(4) Turn the revolving nosepiece until the 4× (scanning objective) snaps into place below the body tube.

(5) Use the knob below the stage so that the specimen in the slide is positioned below the objective.

(6) Look into the ocular and continue to move the slide until you are able to see a blurry version of the specimen. This takes patience! Please go slowly.

*** *Tip*: If you are having difficulty, try to find the edge of the coverslip on the slide first and then, move to the specimen. Also, the smaller the organism, the more difficult it will be to find. Try working with the larger cells first and leave the mixed bacteria slide for last! ***

(7) Use the coarse adjustment knob to begin to focus on the specimen (this is the *only* time you should need to use the coarse adjustment knob).

(8) Use the fine adjustment knob to fine focus on the specimen.

(9) Switch to the next objective (10×).

(10) Fine focus the specimen and position it in the center of the viewing field.

(11) Switch to the next objective (40×), if needed.

(12) Fine focus the specimen and position it in the center of the viewing field.

When viewing each of the slides, note any unique characteristics you observe. Compare the different cell types in the slides. Do they all look the same or different? What similarities do you see? What differences do you see? Draw pictures of what you observe. You will need this information for your post-laboratory thinking questions.

Sterile Technique: Beginning soon and for the rest of this laboratory course, you will be working with the bacterium, *Serratia marcescens*. Although *S. marcescens* is ubiquitous, you need to ensure that the cultures you work with do not get contaminated with other organisms as you work. To do this, you will use sterile technique. Contamination of your *S. marcescens* cultures with other organisms will skew your results. From this week forward, when you work with cultures of bacteria, you should follow the steps listed below to decrease the possibility of contamination and to ensure that you minimize the amount of organism you carry with you when you leave the laboratory:

(1) Wash your hands upon entering the laboratory.

(2) Use 70% ethanol to clean off your laboratory bench surface.

(3) Set up a Bunsen burner and ignite it.

(4) Only open tubes of organisms when you need them and open them up close to the flame. Your instructor will demonstrate how to work with the organism near the flame.

(5) At the close of lab, put your Bunsen burner away.

(6) Wipe down your laboratory bench surface with 70% ethanol.

(7) Wash your hands before you leave the laboratory.

Check out S. marcescens under the microscope! Because you will be working with *S. marcescens* for a vast majority of the course, it is important that you appreciate the size of the organism and how to work with it. As such, this week, you will inoculate a slide with *S. marcescens* so that you can stain it to look at it under the microscope.

Contrast, as discussed earlier, is required because *S. marcescens* is transparent. You will use a dye called methylene blue for the staining technique. To inoculate *S. marcescens* onto a slide, follow the procedure below (your instructor will demonstrate this technique):

(1) Get your workstation ready by clearing the area, washing the lab bench down with 70% ethanol, getting your Bunsen burner ready, and retrieving the supplies/reagents you will need for this lab exercise.

(2) Obtain a microscope slide and pass it through the Bunsen burner flame a few times to remove the oils from the slide. Use a test tube holder or clothespin to do this (*do not touch the slide*—you will get burned!).

(3) Sterilize an inoculating loop until it is red hot using the Bunsen burner flame.

(4) Let the loop cool (but hold it near the flame).

(5) Open up the lid of the culture tube and pass the top of the tube through the flame.

(6) Put the sterilized loop into the culture and ensure that you obtain a bubble in the loop.

(7) Pass the top of the culture tube through the flame and close the lid.

(8) Touch the loop to the microscope slide and spread out the culture fluid onto the slide.

(9) Flame the loop in the Bunsen burner flame to sterilize it.

(10) Allow the fluid on the slide to air-dry.

(11) Once the fluid is dry on the slide, pass the slide (with the inoculated side facing up) through the flame to heat fix the organisms to the slide. Heat fixing kills the organisms and allows them to adhere to the slide so that they are not washed off during the staining process.

To stain the bacteria on the slide, perform the following technique (your instructor will demonstrate this technique):

(1) Place the slide on a staining tray.

(2) Flood the slide with methylene blue.

(3) Wait 1 min.

(4) Flood the slide with water to remove the excess methylene blue.

(5) Place the slide between the pages of bibulous paper and close the book. Pat the book gently to remove the excess water.

(6) Obtain a microscope and use the directions above to find the organism at 40× objective. What do you see? How does the size of the organism compare to the sizes of the cells that you looked at during the beginning of this laboratory session? You will need this information for your post-laboratory thinking questions.

When you are finished using your microscope, *always* return the objective over the condenser back to scanning and clean off all lenses.

If you own your own laptop, please bring it to lab next session. If you do not, please bring a device to save files.

NAME ______________________________ *DATE* ____________

Exercise 2C: Using Microscopy to Evaluate Cell Size and Complexity Post-laboratory Thinking Questions

Directions

Answer the following questions upon completion of the laboratory exercises:

(1) For the first part of the laboratory, you were asked to look at five prepared slides and to draw what you saw for each of the slides. Reproduce your drawings in the circles in Figure 2.3 and label any cellular components that you are able to identify (e.g., intracellular organelles, external features such as flagella/cilia, etc.—this will require that you perform additional research beyond what was discussed in the lab).

(2) During the laboratory, you were also asked to jot down your observations about each of the specimens you observed under the microscope (five prepared slides, one *S. marcescens* slide). Specifically, you were asked to note any differences and similarities between each of the specimens. Please summarize your observations.

(3) Perform some online research to determine the average diameter of each of the cell types you looked at in the lab and complete Table 2.2. Please note the units requested in the Table, um. You may have to use metric conversion to obtain the correct units.

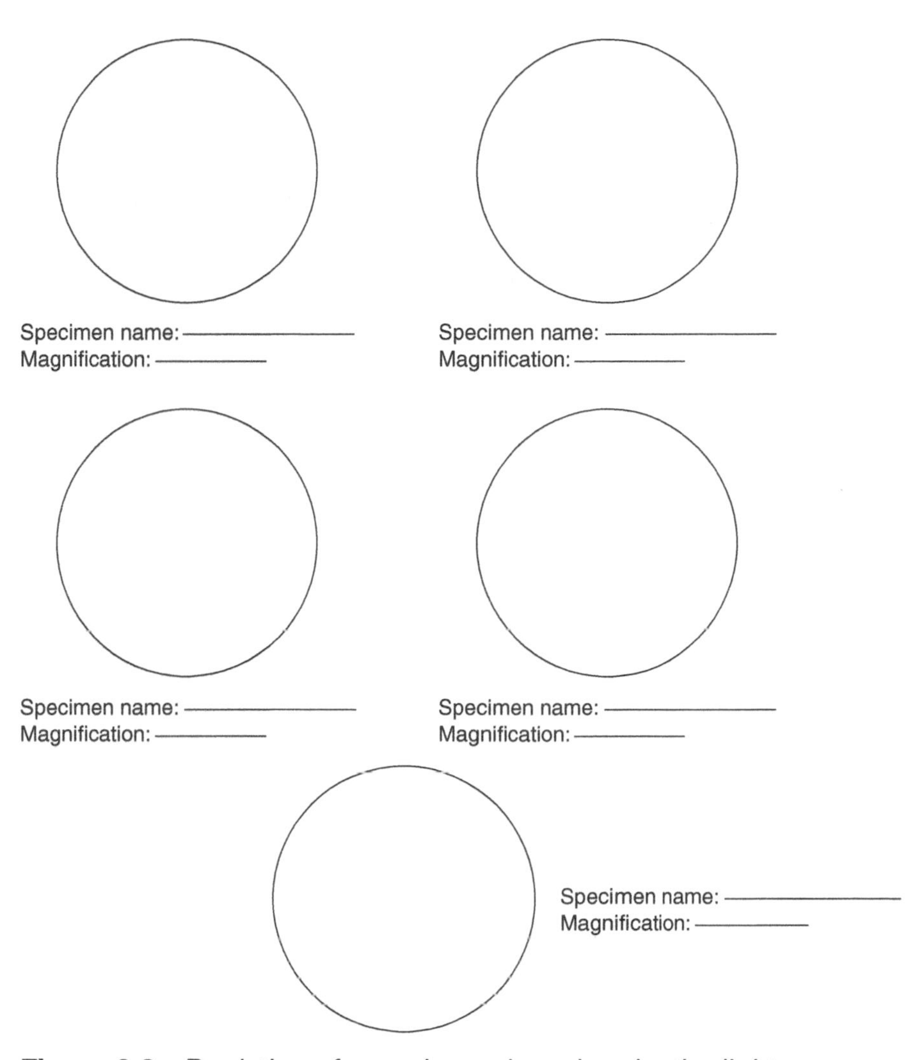

Figure 2.3 Depiction of organisms viewed under the light microscope.

Table 2.2 Average diameter (um) of cells viewed in this exercise.

Slide	Average diameter (um)
Prepared slide of mixed bacteria	
Prepared slide of *Paramecium*	
Prepared slide of *Saccharomyces cerevisiae*	
Prepared slide of an onion root tip	
Prepared slide of human skin cells	
S. marcescens	

EXERCISE 2C

(4) What observations can you make with respect to cell size and complexity based upon your answers to questions 1–3 above?

(5) Compare the cell size you determined for *S. marcescens* to the cell size you determined for the mixed bacteria prepared slide. What conclusions can you make about the size of *S. marcescens* compared to the mixed bacteria? Support your conclusions with facts.

NAME ______________________________ *DATE* ____________

Exercise 3A: The Bacterial Growth Curve Pre-laboratory Thinking Questions

Directions

Read over the introduction and protocols for this laboratory exercise and answer the following questions to ensure that you are prepared for the session:

(1) What are the objectives for today's laboratory (provide a numbered list)?

(2) In your own words, describe the characteristics of the four stages of the bacterial growth curve.

(3) How does a spectrophotometer work and why can it be used to monitor bacterial growth?

(4) As part of this laboratory, you will learn how to use Microsoft Excel to prepare graphs. Why is it important to become proficient in preparing and interpreting graphs for scientists?

The Fundamentals of Scientific Research: An Introductory Laboratory Manual,
First Edition. Marcy A. Kelly.

Companion website: www.wiley.com\go\kelly\fundamentals

NAME ______________________________ *DATE* ____________

Exercise 3B: The Bacterial Growth Curve

Introduction

Studying the Growth of Bacteria: Bacterial growth is due to an increase in the number of bacterial cells present, not an increase in their cell size. Bacterial growth occurs through a process called binary fission. Binary fission is an asexual mode of reproduction in which a single bacterial cell separates into two identical daughter cells. The bacterial cell cycle follows a single bacterium from its formation to its division by binary fission. This cycle proceeds as depicted in Figure 3.1.

Because bacterial growth is due to an increase in the number of bacteria within a population, scientists evaluate bacterial growth by studying the bacterial growth curve in a microbial culture. Microbial cultures are simply populations of bacteria in a test tube. The bacterial growth curve has four phases that can be appreciated in Figure 3.2. Please note that the *y*-axis of the graph uses a logarithmic scale because optimal bacterial growth is exponential (1 bacterium makes 2 bacteria, which make 4 bacteria, which make 8 bacteria, 16, 32, 64, etc.).

The hallmarks of bacterial cells in the four different phases of the growth curve are as follows:

(1) Lag phase—The bacteria have just been introduced into fresh culture media. There is no immediate increase in cell growth, but rather the cells are getting ready to grow. They are "checking out" their new environment and synthesizing the necessary biological molecules and enzymes that they will need for growth.

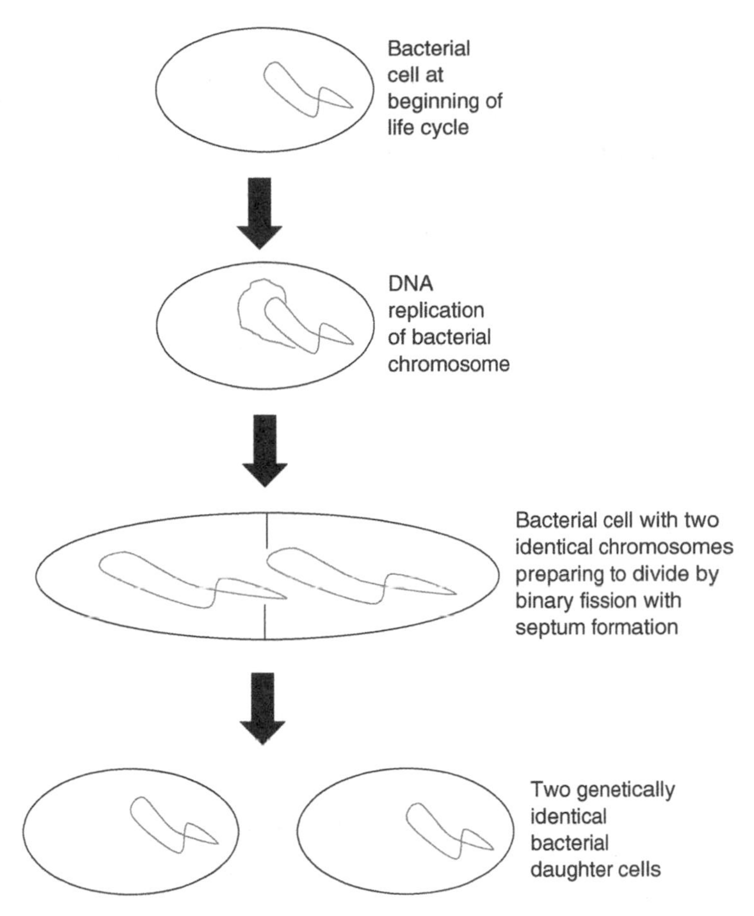

Figure 3.1 Bacterial cell cycle; binary fission. (*See insert for color representation of the figure.*)

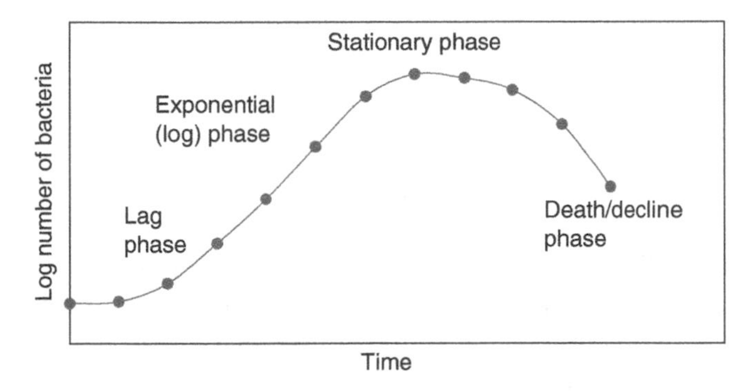

Figure 3.2 The bacterial growth curve. (*See insert for color representation of the figure.*)

(2) Log or exponential phase—The bacteria are ready to grow optimally. Nutrients are plentiful, there is enough oxygen present (if they are aerobic organisms like *Serratia marcescens*), and the bacteria are actively metabolizing. During log phase, each bacterium has a particular doubling time. This is the amount of time from the generation of a new cell to that cells own division. *S. marcescens* has a doubling time of approximately 45 min. *Escherichia coli* has a doubling time of approximately 20 min. *Mycobacterium tuberculosis* has a doubling time of approximately 24 h. Therefore, bacterial doubling times can be vastly different, depending upon the organism. We will work with *E. coli* today so that you may practice proper sterile technique and so that you will be able to appreciate the entire bacterial growth curve in a single laboratory session. When we begin working with *S. marcescens*, you will not be able to evaluate the entire growth curve in a single laboratory session because of the longer doubling time of the organism.

(3) Stationary phase—As you can see from the growth curve, when bacteria enter stationary phase, it appears that bacterial growth ceases. In reality, during stationary phase, bacterial growth rates are equivalent to bacterial death rates. This occurs because the nutrients in the media have become limited, metabolic by-products (wastes) have accumulated to toxic levels in the media, and there is a decrease in the solubility of oxygen in the media. Some bacteria are able to enter a dormant state upon entry into stationary phase in order to protect themselves from bacterial cell death. These bacteria typically form spores. *Bacillus anthracis* is an example of a bacterium that is able to form spores under these conditions. Once the environment is more favorable, the organisms shed their spores and begin growing again. *S. marcescens* is not able to form spores.

(4) Death or decline phase—At this stage of the bacterial cell cycle, the bacteria are completely starved and depleted of oxygen. In addition, the accumulation of toxic materials in the media has become deleterious for the bacteria. As such, bacterial death increases dramatically.

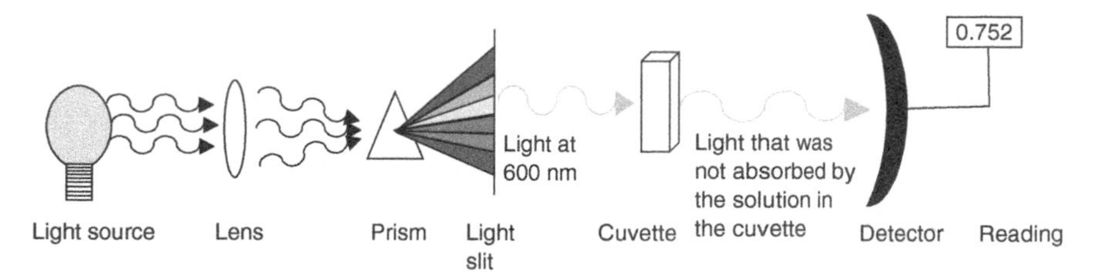

Figure 3.3 The functional components of a spectrophotometer. (*See insert for color representation of the figure.*)

Spectrophotometry: As mentioned earlier, there is instrumentation that can help scientists evaluate which stage of the bacterial growth curve cells grown in culture are in. This piece of equipment is called a spectrophotometer. Spectrophotometers are instruments designed to determine the amount of light that is absorbed by a solution. The more dilute a solution, the less light that is absorbed (water absorbs very little light). The more concentrated a solution, the more light that is absorbed (orange juice absorbs a lot of light). With respect to bacterial cell cultures, the more bacteria present, the denser the culture, and the more light is absorbed.

A spectrophotometer has many components that work together to provide absorbance values for samples (in our case, the bacterial cultures). They are depicted in Figure 3.3.

Basically, a light source is collected by a lens system and separated out into the different wavelengths of light using a prism. The wavelength of light that you wish to monitor is shined on the sample, and a detector detects how much of that light at the wavelength you selected can pass through the sample solution without being absorbed. The spectrophotometer provides you with a numerical reading to represent the amount of light that is absorbed by the sample. The units obtained by measuring solutions using a spectrophotometer are called absorbance units. Typically, spectrophotometers used to measure absorbance units run on a scale of 0 absorbance units to 2.0 absorbance units.

EXERCISE 3B

Spectrophotometry follows the principles of the Lambert–Beer law. This law states that absorbance units are directly and linearly related to the concentration of a solution. This means that the higher the absorbance units, the more light is absorbed by the solution and the more concentrated the solution.

When measuring a sample using the spectrophotometer, you must first set a blank to serve as a control. In our case, for this exercise, the blank will be media without *E. coli* in it. Once the blank is set (i.e., the spectrophotometer reads the blank as 0 absorbance units), you may begin measuring your samples.

The Bacterial Growth Curve and Spectrophotometry: A vast majority of microbiological research laboratories require bacterial growth cultures that are midway through the log phase (aka mid-log) of the growth curve for their work. In order to determine the growth phase of their cultures, scientists have come up with the following absorbance readings, when a spectrophotometer is set to read at 600 nm, to represent the different phases of the growth curve (Table 3.1).

Absorbance readings for bacterial cultures are usually reported as optical densities (ODs) because the absorbance of light is related to the density of the culture.

Table 3.1 Approximate optical densities (ODs) at 600 nm for bacterial cultures at different phases of the bacterial growth curve.

Growth phases	Optical density at 600 nm (OD)
Lag phase	0–0.25
Log phase	0.25–0.8
Stationary phase	0.8–1.8
Death phase	Cannot be evaluated using a spectrophotometer. When bacteria die, they still remain the same structurally in most cases, so the spectrophotometer readings are similar to those of stationary phase

In addition to the information in Table 3.1, in general, a culture with an OD of 1.0 contains approximately 1×10^9 bacterial/ml. Haddix *et al.* (2000) went a step further and demonstrated that for *S. marcescens*, a culture with an OD of 1.0 contains approximately 5.80×10^8 *S. marcescens*/ml. You will use this number in future exercises for this laboratory.

Let's Put This to Practice!

Prior to the start of this laboratory session, stationary phase cultures of *E. coli* were diluted down to create cultures in lag phase of the bacterial growth curve (OD = 0.2). You are going to work in groups of 4 to monitor the growth of one of these cultures using the spectrophotometer for 2 h (120 min). You will grow your organisms in a shaking incubator set to 37°C for this study. At the end of the 2 h period, you will be able to create a growth curve for the organism. Once you obtain your data, you will graph that data using Microsoft Excel and use that information to determine an approximate doubling time for the organism.

There will be considerable down time during this laboratory between each time point. During this time, you will learn how to graph using Microsoft Excel. The protocol describing how to graph using Microsoft Excel follows the growth curve protocol:

(1) Turn on the spectrophotometer to allow it to warm up.

(2) Get your workstation ready by clearing the area, washing the lab bench down with 70% ethanol, getting your Bunsen burner ready, and retrieving the supplies/reagents you will need for the lab exercise.

(3) Use sterile technique to remove 2 ml of fresh, sterile media into a spectrophotometer cuvette. You will use the fresh media as a blank for the spectrophotometer. The protocol for this is as follows (your laboratory instructor will demonstrate this technique):

(a) Open the media container near the Bunsen burner flame.

(b) Quickly flame the lid of the media container.

(c) Using a sterile pipette, obtain 2 ml of the media (be sure not to touch the tip of the pipette to any surfaces excluding the media!).

(d) Pipette the media into the cuvette. Do this in close proximity to the Bunsen burner flame.

(e) Flame the lid of the container.

(f) Close the container lid.

(g) Place the cuvette on the lab bench in close proximity to the Bunsen burner flame.

(4) Use sterile technique to remove 2 ml of your culture into a spectrophotometer cuvette following steps (a–g). Once you remove the 2 ml, make sure to return the culture to the 37°C shaking incubator.

(5) Set up the spectrophotometer by ensuring that it will take readings in absorbance units (A) and change the wavelength to 600 nm. Your instructor will show you how to do this.

(6) Place your blank cuvette into the spectrophotometer, ensuring that the clear sides of the cuvette are in line with the light path. You instructor will show you how to do this.

(7) Close the lid on the spectrophotometer and hit the 0 button to blank the spectrophotometer. Your instructor will show you how to do this. When the machine is blanked, you will see a "0" reading on the view screen. If you do not see a "0," hit the blank button again and wait to see if you get a "0" reading. If you still have trouble, please ask for assistance.

(8) Open the lid, remove your blank, and discard it in the red biohazard waste.

(9) Place your culture sample into the machine, ensuring that the clear sides of the cuvette are in line with the light path.

(10) Close the lid and record the reading on the view screen. You do not need to press any buttons on the machine to take a reading. The reading is the optical density for your culture at 600 nm.

(11) Open the lid, remove your sample, and discard it in the red biohazard waste.

(12) Repeat steps 3–11 every half hour for 2 h (120 min; you should end up with five readings).

(13) Record the readings in Table 3.2.

(14) Use Excel to create a growth curve of the data in the last two columns of Table 3.2.

EXERCISE 3B

Table 3.2 Growth of *E. coli* at OD_{600}.

Time (min)	Optical density at 600 nm (OD)
0	
30	
60	
90	
120	

Graphing Using Microsoft Excel: Experimental data is often reported in graphical format. As such, it is important that you become proficient in using graphing programs, such as Microsoft Excel. Please review the following directions and work on the sample problems that follow during the down time in the laboratory. Your instructor will ensure that you have completed the sample problems accurately and answer any questions you might have.

Please note that these instructions are for Microsoft Excel 2010 on a PC. Generally, the different versions of Excel are similar, but there may be some slight differences.

To open Excel:

(1) Double-click on the Microsoft Excel icon located on the desktop. If there is no Microsoft Excel icon on the desktop, follow the directions below:

(a) Click on start at the bottom left of the screen.

(b) Click on All Programs.

(c) Click on Microsoft Office.

(d) Click on Microsoft Excel.

To create your data table (using the Excel spreadsheet):

(1) Type in the heading for the X axis in the A1 box. The X axis on a graph should display the independent variables or constant values (with their units) such as concentration, time, temperature, or pH. If heading is too long to fit in the A1 box, you may increase the box width by placing the cursor on the line between the A and B box at the top of the spreadsheet so you get a double-headed arrow and dragging to the width you want.

(2) Type in the heading for the Y axis in the B1 box. The Y axis on a graph should display the dependent variable values (with their units) such as absorbance. Again, if heading is too long, drag line between columns B and C as described above.

(3) Enter the values for the X axis (numbers only, no letters or symbols) in the A column under your X axis heading (i.e., A2, A3, A4, etc.). A degree (°) symbol can be added after temperature values. To do this, hold down the Alt key and type 248 if you're using a PC or hold down the Alt key and the Shift key and hit the number 8 key if you're using a Mac after you type in the temperature number.

(4) Enter the values for the Y axis (numbers only, no letters or symbols) in the B column under your Y axis heading (i.e., B2, B3, B4, etc.).

Making your graph:

(1) Use the mouse to select the entire table that you just created (be sure to include the headings that you typed in boxes A1 and B1).

(2) Click on the Insert menu tab located in the top menu bar.

(3) In the Chart dialogue box, select the scatter icon. Then select the top left scatter chart shown (aka scatter chart with only markers).

(4) Your chart will now be displayed on top of your spreadsheet. To enlarge your chart, place the mouse on the top left corner, and drag it diagonally. Once you enlarge the chart to a better size for you to work with, move the chart over (by dragging it) to the right so that you can see your data on the spreadsheet.

(5) You will need to change the title of your chart. To do so, double-click the default title that is on your chart until the cursor appears. Delete the default title and type in a newer, more appropriate title. You can also change the title using the Layout menu tab (see below).

To add titles to your axes:

(1) Click on the Layout menu tab located in the top menu bar.

(2) In the Layout dialogue box, select the axis title icon.

(3) To add a title to your x (horizontal) axis, select the primary horizontal axis title icon and then pick the title below axis selection. A textbox will then appear on the horizontal axis of your graph. Type your x-axis title and units in the textbox.

(4) To add a title to your y (vertical) axis, select the primary vertical axis title icon and then pick the rotated title selection. A textbox will then appear on the vertical axis of your graph. Type your y-axis title and units in the textbox.

To add a trendline and formulas to your graph:

(1) Click on the Layout menu tab in the top menu bar.

(2) In the Analysis dialogue box, select the trendline icon.

(3) On the scroll down menu, select more trendline options (the last selection on the scroll down menu).

(4) In the Format Trendline box that appears, make sure that:

(a) Under Trend/Regression types, linear is selected.

(b) Under Trendline name, automatic is selected.

(c) Under Forecast, forward = 0.0 periods and backwards = 0.0 periods.

(d) Then select display equation on chart and display R-squared value on chart.

Depending upon the type of data, you might also have to select set intercept to 0. This sets the y-intercept at zero so when your formula ($y = mx + b$) comes up, the b-value will be zero and not appear. You would need to do this if your data includes a 0 x-value for a 0 y-value. For this laboratory course, you would only need to do this when you generate standard curves (Exercises 4 and 10).

A review of algebra: $y = mx + b$ is a formula used to identify the slope and y-intercept for any given x- and y-values on a graph. For the equation, y is the y-value at a given x-value, m is the slope of the line (you will need to use this value for data analysis as it indicates the rate of reactions), x is the x-value at a given y-value, and b is the y-intercept (the y-value at which the plot crosses the Y axis).

(5) Once you have the trendline and formulas on your chart, move the formulas to the right of the plot:

(a) Click on the formula—a border should appear around the formula.

(b) Click on the border and drag it to the right of the plot.

Determining the correlation coefficient (r):

The correlation coefficient indicates how closely the points on your graph fit a straight line. If your correlation coefficient is not close to 1, your data is not linear:

(1) On a piece of paper, write down the R-squared value.

(2) Click in any empty box on your spreadsheet—a heavy border should appear around the box.

(3) Type in =sqrt (the R-squared value you wrote down).

For example, if your R-squared value was 0.949, then type =sqrt(0.949).

(4) Hit Enter.

(5) Write down the number that appears in the box. This is the correlation coefficient (r).

Adding the correlation coefficient to the graph:

(1) Click on the formulas on your graph—a box should appear around the formulas.

(2) Move the mouse into the hatched box and click on it again—a cursor should appear.

(3) Use the arrow buttons on your keyboard to position the cursor after the R-squared value.

(4) Hit Enter so the cursor is below the R-squared value.

(5) Type in $r =$ the number you determined on the spreadsheet. For example, if you determined that the r-value was 0.949, then type in $r = 0.949$.

(6) Click anywhere on the graph (except in the formula box) to remove the cursor from the formula box.

Using the $y = mx + b$ formula to solve for an x-value when you know the y-value:

You would use this to determine the concentration of an unknown solution (x) at a given absorbance (y; This will be needed for Exercises 4 and 10):

(1) Write down the $y = mx + b$ equation on a piece of paper.

(2) Click in any empty box on your spreadsheet—a heavy border should appear around the box.

(3) Type in the known y-value.

(4) Click on the box next to the one you just typed in—a heavy border should appear around the box.

(5) Type in =(the letter and number of the column and row of the y-value you just entered—the b-value from the $y = mx + b$ equation)/ the m-value from the $y = mx + b$ equation. This can be written another way: = (letter# – b)/m. For example, suppose the $y = mx + b$ equation on your graph is $y = 29.524 + 5$, you would then type in = (D11 – 5)/29.524. If you set the y-intercept to 0 and your $y = mx + b$ equation on your graph is $y = 29.524$, you would then type in = (D11)/29.524.

(6) Hit Enter—the x-value (at that y) will appear in that box.

Making a multi-line graph:

You would use the following protocol to chart several lines on the same graph—for example, if you wanted to chart the effect of different temperatures on bacterial growth:

(1) Type in a heading for the X axis data in the A1 box, for example, Time (seconds).

(2) Type in a heading for the first set of Y axis data in the B1 box, for example, 4°C.

(3) Type in a heading for the second set of Y axis data in the C1 box, for example, 23°C.

(4) Type in a heading for the third set of *Y* axis data in the D1 box, for example, 37°C. And so on.

(5) Type in your *X* and *Y* axes data in the appropriate columns.

(6) Select the table and follow the directions as described above.

(7) Each line on the graph should have its own set of formulas ($y = mx + b$, R-squared value, and r-value). All of the formulas should be to the right of the plot and labeled appropriately (get the cursor in the formula box as described above and type in the labels above the formulas).

Sample Problems for Graphing on Excel

Graph with a Single Line: Systolic blood pressure was taken on a number of human males of varying ages resulting in the data shown in Table 3.3.

Table 3.3 Systolic blood pressures of nine human males of varying ages.

Subject	Age in years	Blood pressure (mm Hg)
1	19	122
2	25	125
3	30	126
4	42	129
5	46	130
6	52	135
7	57	138
8	62	142
9	70	145

How would you enter the above data in a table in Excel? (Remember that for a single-line graph, you only want two columns of data, not three, so you must pick which two of the three columns above should be used in the spreadsheet.) Once you have decided how to set up the data, make a table and follow the directions in the Excel handout to construct a single-line graph. Do not set the y-intercept to 0.

Now, calculate the predicted age for a male with a blood pressure of 133 (solving for x when you know y).

Multi-line Graph: Microbiologists discovered a species of archaebacterium growing in a hot spring in Yellowstone National Park. They isolated an enzyme from this archaebacterium that produces a pink pigment. In order to determine the optimal temperature for this enzyme, they set up the following experiment.

The microbiologists mixed the appropriate substrate with the enzyme in reaction tubes and incubated them at 22, 45, or 84°C for 10, 30, and 60 min. At the end of the specified time intervals, they measured the absorbance of the tubes using a spectrophotometer. Their results appear in Table 3.4.

Table 3.4 Absorbance of pink pigment produced by an archaeal enzyme at different temperatures.

Time (min)	Absorbance at 21°C	Absorbance at 42°C	Absorbance at 84°C
10	0.115	0.280	0.120
30	0.315	0.675	0.120
60	0.600	0.980	0.120

Make a multi-line graph of the data in the above table. Note that no pigment was produced at time 0; therefore, set your y-intercept to 0. At which temperature is the rate of pigment production greatest? Rate is the slope of the line (the m in the $y = mx + b$ equation).

Answer to Excel Sample Problems
Single-line graph (Table 3.5 and Figure 3.4):

Table 3.5 Data for sample problem one as input into Excel.

Age in years	Blood pressure (mmHg)
19	122
25	125
30	126
42	129
46	130
52	135
57	138
62	142
70	145
r-value	0.977701386
Calculated unknown age	*Blood pressure at unknown age*
46.01513802	133

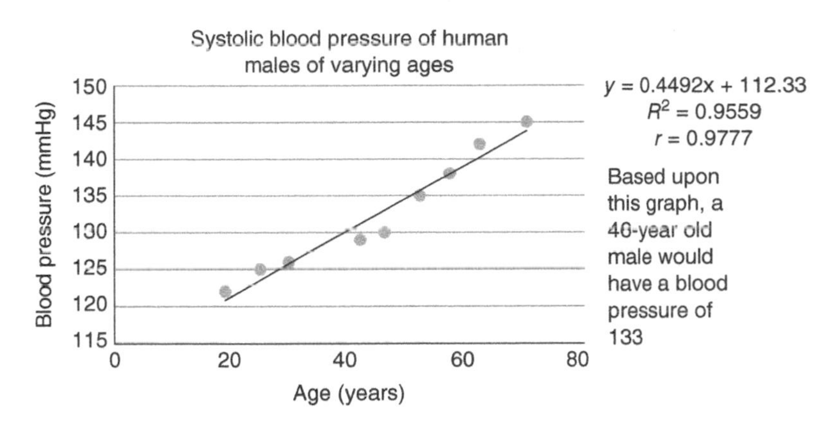

Figure 3.4 Blood pressure Excel sample problem graph. (*See insert for color representation of the figure.*)

Multi-line graph (Table 3.6 and Figure 3.5):

Table 3.6 Data for sample problem two as input into Excel.

Time (min)	Absorbance at 21°C	Absorbance at 42°C	Absorbance at 84°C
10	0.115	0.28	0.12
30	0.315	0.675	0.12
60	0.6	0.98	0.12
r-values	0.998448797	0.919565115	NA

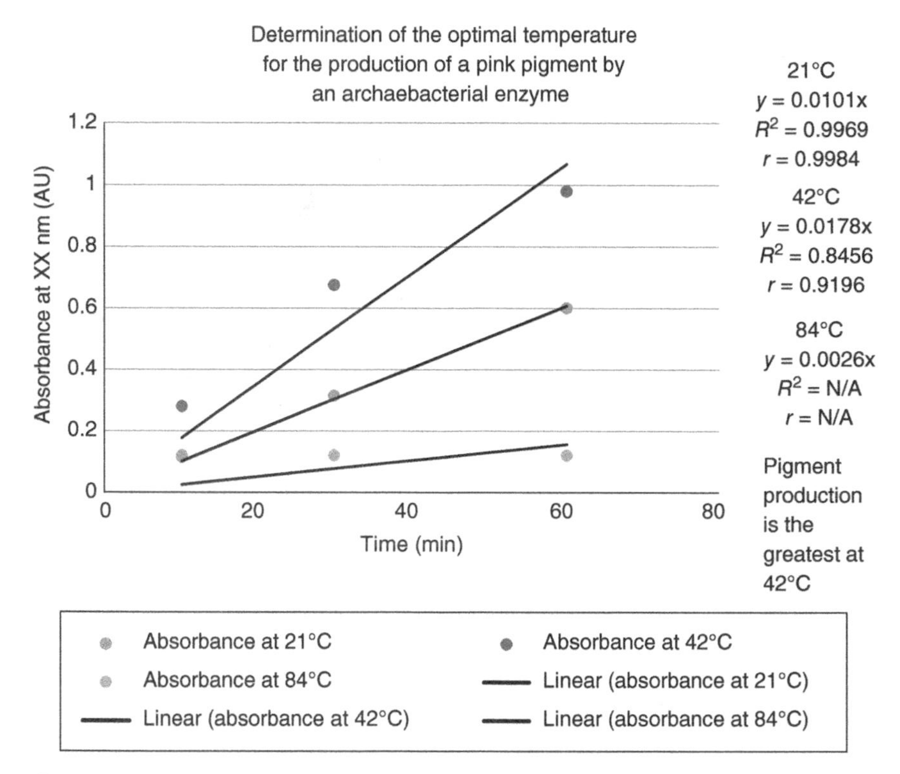

Figure 3.5 Archaeal pigment Excel sample problem graph. (*See insert for color representation of the figure.*)

If you own your own laptop, please bring it to lab next session. If you do not, please bring a device to save files.

NAME ______________________________ *DATE* ____________

Exercise 3C: The Bacterial Growth Curve Post-laboratory Thinking Questions

Directions

Answer the following questions upon completion of the laboratory exercises:

For the first part of the laboratory, you were asked to use the spectrophotometer to measure the absorbance of your bacterial culture for 2h (120min) at 600nm. At the end of the laboratory session, you were asked to record the data you obtained in Table 3.2. Please ensure that you have all of the information in Table 3.2 to answer the following questions:

(1) Please use Microsoft Excel to prepare a graph of the data in Table 3.2. Copy/paste your graph into this worksheet.

(2) In your own words, summarize your findings from your graph. Be sure to specifically refer to the information presented in your graph and the information presented in the introduction for this laboratory.

MODULE 2

Working with and Learning About *Serratia marcescens* in the Laboratory

NAME ______________________________ *DATE* ____________

Exercise 4A: Protein Concentration Versus Growth Stage Pre-laboratory Thinking Questions

Directions

Read over the introduction and protocols for this laboratory exercise and answer the following questions to ensure that you are prepared for the session:

(1) What are the objectives for today's laboratory (provide a numbered list)?

(2) Proper protein folding is essential for the activity of a protein. In your own words, please discuss the details of protein folding.

(3) What is the difference between a qualitative and quantitative test?

(4) How does the Lowry assay work?

(5) What relationship do you expect to find between growth stage and protein concentration (this is asking you to develop a hypothesis for this exercise)? Support your answer.

The Fundamentals of Scientific Research: An Introductory Laboratory Manual, First Edition. Marcy A. Kelly.

Companion website: www.wiley.com\go\kelly\fundamentals

NAME ______________________________ *DATE* ____________

Exercise 4B: Protein Concentration Versus Growth Stage

Introduction

Complex Biological Molecules: All cells, including the cells of bacteria such as *S. marcescens*, are composed of four classes of complex biological molecules: proteins, carbohydrates, lipids (fats), and nucleic acids. These four groups of complex biological molecules are composed of hydrocarbon chains linked to functional groups. The functional groups of the different complex biological molecules give the biological molecules their chemical properties (or reactivity).

The four different complex biological molecules are composed of repeating subunits called monomers. The monomers are linked together by dehydration synthesis reactions to form the complex biological molecules. As the name implies, dehydration reactions result in the generation of water molecules. Because the complex biological molecules are composed of many monomers covalently linked together, they are considered polymers. A dehydration reaction is illustrated in Figure 4.1.

As described earlier, each of the classes of complex biological molecules has different chemical characteristics. These different chemical characteristics impact both the structure and function of the biological molecules within the cell. For this laboratory exercise, we will focus on proteins, but it is expected that you will realize the importance of the other biological molecules within the cell.

Proteins: Proteins are the most abundant complex biological molecule within a cell, and they have many functions within cells. For

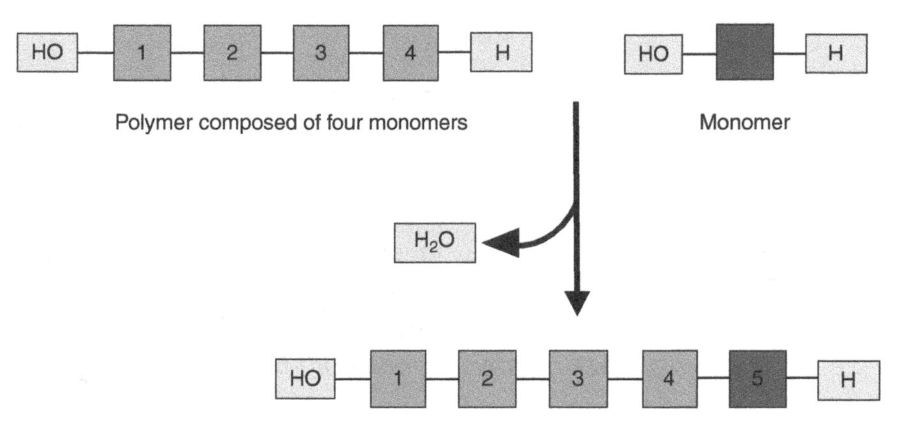

Figure 4.1 Dehydration synthesis between a polymer of four monomers and a single monomer to yield a polymer of five monomers. A single molecule of water is lost during the reaction. (*See insert for color representation of the figure.*)

example, they can serve as structural molecules responsible for maintaining cellular shape. They can also serve as functional molecules. Enzymes are catalytic proteins responsible for carrying out the metabolic reactions within a cell. Other examples of functional proteins include the proteins that serve as hormones, contractile proteins of the muscle, and defense proteins of the immune response.

The monomers of proteins are the amino acids. There are 20 amino acids. The amino acids consist of an amino group, a carboxyl group, and a variable group or R-group bonded to a central carbon. The basic amino acid structure is depicted in Figure 4.2. The R-group gives each amino acid its own chemical properties. The composition of the different amino acids in a given protein can give that protein or specific regions within that protein acidic, basic, polar, nonpolar, or neutral characteristics.

Proper protein functioning requires that the protein is folded correctly when it is synthesized by the cell. During protein synthesis, the different amino acids are joined together by dehydration

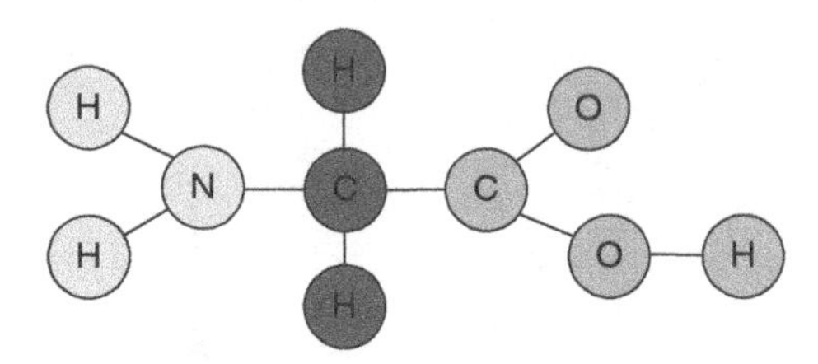

Figure 4.2 An amino acid. Illustration includes the amino group (yellow), central or alpha carbon, and associated R-group which is variable in each amino acid (red), and the carboxyl group (blue). (*See insert for color representation of the figure.*)

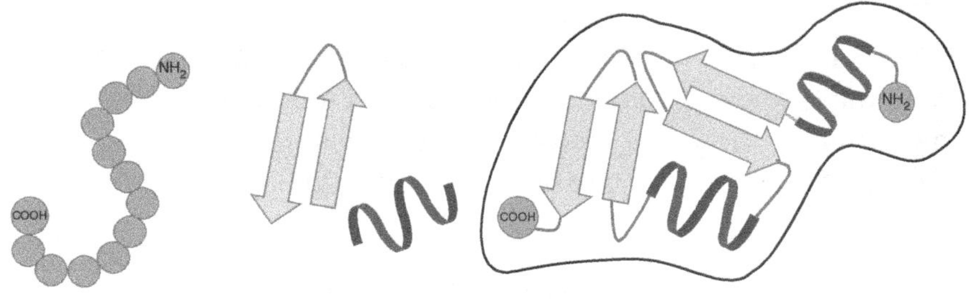

Figure 4.3 Protein folding. The primary structure of a protein consists of a linear chain of amino acids (blue circles). The secondary structure consists of beta pleated sheets (yellow arrows) and alpha helices (red spirals). The tertiary structure is a globular functional protein consisting of secondary structures folded in a specific, coordinated fashion to ensure that protein structure enables protein functioning. (*See insert for color representation of the figure.*)

synthesis reactions to form peptide bonds. A long chain of amino acids linked together by the peptide bonds represents a protein in its primary structure. The protein in its primary structure is folded within the cell by a highly coordinated series of events. Figure 4.3 depicts these events. The primary structure is first twisted into alpha helices and beta pleated sheets. Both of these structures represent the secondary structure of proteins. They are formed due to the formation of hydrogen bonds between different amino acids in the primary structure. After the secondary structure has been stabilized, the protein will be folded into its tertiary structure. The tertiary structure is stabilized by various types of bonds between the R-groups of the different amino acids in the protein. These bonds include hydrogen bonds, ionic bonds, disulfide bonds, Van der Waals interactions, and hydrophobic interactions. The strongest of the bonds that stabilize the tertiary structure of proteins are the disulfide bonds. These bonds are covalent bonds between two cysteine amino acids that are adjacent to each other in the protein. Cysteines have sulfhydryl groups as their R-groups. Disulfide bonds form between two sulfur atoms present in the adjacent cysteines. Tertiary structure proteins are considered globular proteins. Some proteins require a forth level of protein folding for proper functioning. These proteins are folded into the quaternary structure. Quaternary structure proteins are simply several globular tertiary structure proteins that are bonded together. Hemoglobin, the oxygen-carrying molecule in human blood, is a quaternary protein composed of four globular protein subunits bonded together.

After folding, some proteins require additional modifications in order to become functionally active. These modifications can include the addition of a carbohydrate group or a lipid group or the removal of the first few amino acids in the protein (which typically serve to aid in the localization of that protein to a specific region within the cell).

Measuring Protein within Cells: Several techniques have been developed to measure protein concentration within cells. These techniques include both qualitative and quantitative techniques. Qualitative techniques enable scientists to determine if a molecule (such as protein) is present or not, but they do not provide a numerical amount or concentration for that molecule. Quantitative techniques enable scientists to determine a numerical concentration for the molecule being tested. The addition of biuret reagent to a sample would enable scientists to determine if protein was present in that sample by a color change from blue to purple. This is an example of a qualitative test because there is no numerical value associated with the test result. For a more accurate estimate of protein concentration within a cell, techniques such as the Bradford assay or the Lowry assay are used. These assays use the spectrophotometer to generate a numerical value representing protein concentration within a sample.

For this laboratory exercise, we are going to use the Lowry assay to evaluate the amount of protein present in *S. marcescens* at different stages of its growth curve. The Lowry assay was developed by Oliver Lowry in the 1951. For the assay, two solutions are added to a sample, and if protein is present in that sample, there will be a change in color that can be measured quantitatively using a spectrophotometer. The first solution, biuret reagent, contains copper ions. The copper ions interact with peptide bonds and oxidize them. The second reagent, Folin reagent, oxidizes amino acids with R-groups containing aromatic ring structures (tyrosine and tryptophan). As the amino acids are oxidized by the two reagents, the Folin reagent molecule becomes reduced. Reduced Folin reagent changes from yellow to blue. This color change can be measured using the spectrophotometer at a wavelength of 750 nm.

Let's Put This to Practice!

Prior to the start of this laboratory session, stationary-phase cultures of *S. marcescens* were diluted down to different optical densities (OD): 0.1, 0.25, 0.5, 1.0, and 1.5. Each student group will work with one of these cultures to lyse the bacteria. Once lysed, each group will take samples from each of the five lysed cultures for analysis. Following completion of the laboratory, each student will create a graph on Excel to aid in the analysis of the impact of growth stage on protein concentration.

Bacterial Cell Lysis: In order to determine the amount of protein present within your bacterial sample, you must first lyse open the bacterial cells. Lysis is the process by which the bacterial cell wall and cell membrane are broken open, releasing all of the cytoplasmic contents of that cell. To do this, we will expose the bacteria to a detergent, SDS, and high heat. SDS aids in cell lysis because it interrupts lipid–lipid and lipid–protein interactions. These interactions are required to maintain the stability of the cell membrane. If they are interrupted, the cell membrane will break apart.

Lysis Protocol

(1) Spin down your cultures for 10 min at 4600–5000×g.

(2) Pour off the supernatant and resuspend the cells in 4.5 ml TE buffer.

(3) Transfer the solution to a 15 ml conical vial.

(4) Add 500 ul 20% SDS.

(5) Incubate at 95°C for 5 min.

(6) Transfer the solution to an Oak Ridge tube.

(7) Spin down at 15,000×g for 15 min.

(8) Use the supernatant for the Lowry assay as described below.

Lowry Assay

(1) Pipet 100 ul of four protein standards, 100 ul from each lysed culture of *S. marcescens*, and 100 ul of TE buffer into separate 15 ml conical vials (make sure you label each tube so you know which sample you put into them). The TE buffer tube will serve as your blank for the spectrophotometer later in the experiment.

(2) Add 510 ul DC Reagent A to each vial and cap the vials.

(3) Vortex.

(4) Incubate at room temperature for 5 min.

(5) Vortex.

(6) Add 4 ml DC Reagent B to each vial and cap the vials.

(7) Vortex.

(8) Incubate at room temperature for 15 min.

(9) Turn on the spectrophotometer and let it warm up. After it warms up, set it to read absorbances at 750 nm.

(10) Blank the spectrophotometer with 2 ml from the TE sample (pipet into a cuvette and then, blank the spectrophotometer like you did last week).

(11) Pipet 2 ml of each of the samples (the four protein standards and the five *S. marcescens* lysate samples) into cuvettes and take readings. Record your readings in Table 4.1.

Creation of a Standard Curve to Aid in the Determination of the Protein Concentration of Your S. marcescens Sample: In Table 4.1, you might have noticed that you recorded your results in absorbance (A) at 750 nm. You now need to convert the A that you recorded

Table 4.1 Absorbance of protein standards and *S. marcescens* lysates at 750 nm.

Sample	Absorbance at 750 nm (AU)
Protein standard 1 (0.25 mg/ml)	
Protein standard 2 (0.5 mg/ml)	
Protein standard 3 (1.0 mg/ml)	
Protein standard 4 (1.5 mg/ml)	
S. marcescens, OD 0.1	
S. marcescens, OD 0.25	
S. marcescens, OD 0.5	
S. marcescens, OD 1.0	
S. marcescens, OD 1.5	

into protein concentration in mg/ml. To do this, you must first create a standard curve on Microsoft Excel using the data you obtained from the four protein standards (located above the bold line in Table 4.1). Standard curves demonstrate the linear relationship between absorbance at a particular wavelength and concentration. Standard curves can be used to determine the concentration of an unknown in a sample (such as the protein concentration in your *S. marcescens* sample).

Using Microsoft Excel, create a single line graph to illustrate the linear relationship between the concentration of protein in each of your protein standards versus the A you obtained. Make sure you have selected the correct information for your x- and y-axes before you create your graph. Which one is the dependent and which one is your dependent variable? Additionally, when you create your standard curve, remember to set your y-intercept to 0.

After you create your standard curve graph, use the *S. marcescens* A_{750} data in Table 4.1 (located below the bold line in Table 4.1) and the "Using the $y = mx + b$ Formula to Solve for an x Value When You Know the y Value" directions in Exercise 3 of this manual to determine the protein concentration in each of your lysed *S. marcescens* samples. Complete Table 4.2 by listing the protein concentrations you determined.

EXERCISE 4B

Table 4.2 Protein concentration of *S. marcescens* at different phases of the bacterial growth curve.

S. marcescens culture OD_{600}/ growth phase	Protein concentration (mg/ml)
$OD_{600} = 0.1$ (lag phase)	
$OD_{600} = 0.25$ (late lag phase)	
$OD_{600} = 0.5$ (log phase)	
$OD_{600} = 1.0$ (early stationary phase)	
$OD_{600} = 1.5$ (late stationary phase)	

Use the information in Table 4.2 to create a graph using Microsoft Excel to show the relationship between growth phase and protein concentration. You will need this graph for your post-laboratory thinking questions.

If you own your own laptop, please bring it to lab next session. If you do not, please bring a device to save files.

NAME ______________________________ *DATE* ____________

Exercise 4C: Protein Concentration Versus Growth Stage Post-laboratory Thinking Questions

Directions

Answer the following questions upon completion of the laboratory exercises.

For the first part of the laboratory, you were asked to use the spectrophotometer to measure the protein concentration of the protein standards and lysed cultures. You were then asked to record the data you obtained in Table 4.1. Please ensure that you have all of the information in Table 4.1 to answer the following questions:

(1) Please create a standard curve using Microsoft Excel to help you determine the concentration of protein in your *S. marcescens* sample using the data in Table 4.1. Copy/paste the standard curve into this worksheet.

(2) After you created the standard curve, you were asked to use the information in your standard curve to help you determine the amount of protein in the lysed *S. marcescens* samples and record that information in Table 4.2. Reproduce the data you obtained for Table 4.2.

(3) Use the information in Table 4.2 to create a graph using Microsoft Excel to show the relationship between growth phase and protein concentration. Copy/paste that graph into this worksheet.

(4) In your own words, describe the relationship between growth phase and protein concentration using details from your graph to support your answer.

(5) Restate the hypothesis you developed for your pre-laboratory thinking questions. Is your hypothesis correct or incorrect? Use the data presented in the graph you created to answer question 4 to support your answer.

NAME ______________________________ *DATE* _____________

Exercise 5A: Measuring Prodigiosin Pre-laboratory Thinking Questions

Directions

Read over the introduction, protocols, and Haddix and Werner paper (2000) for this laboratory session and answer the following questions to ensure that you are prepared for the session:

(1) What are the objectives for today's laboratory (provide a numbered list)?

(2) In your own words, describe the characteristics of prodigiosin.

(3) What is the maximum wavelength for prodigiosin absorbance?

(4) For this laboratory, you will calculate the units of prodigiosin produced per cell. In your own words, please describe how the equations to calculate units of prodigiosin produced per cell were derived.

(5) What relationship do you expect to find between growth stage and prodigiosin production (this is asking you to develop a hypothesis for this exercise)? Support your answer.

The Fundamentals of Scientific Research: An Introductory Laboratory Manual, First Edition. Marcy A. Kelly.

Companion website: www.wiley.com\go\kelly\fundamentals

NAME ______________________________ *DATE* ____________

Exercise 5B: Measuring Prodigiosin

Introduction

What is prodigiosin? S. marcescens is unique among bacteria because it produces a red cell-associated pigment called prodigiosin. Prodigiosin is a low-molecular weight molecule composed of three pyrrole rings linked together. The exact structure of prodigiosin is depicted in Figure 5.1 (Haddix and Werner, 2000; reviewed in Khanafari *et al.*, 2006).

Prodigiosin is considered a secondary metabolite of *S. marcescens*. Secondary metabolites are molecules that are not directly involved in the growth of an organism. The absence of a secondary metabolite does not impact the survival of the organism that produces it (Haddix and Werner, 2000; reviewed in Khanafari *et al.*, 2006).

Prodigiosin is synthesized as a result of a phenomenon called quorum sensing. Quorum sensing is a method that bacteria use to communicate with each other. In this particular case, *S. marcescens* produces a prodigiosin regulator protein that is secreted into the environment. At low cell numbers, the amount of the regulator protein in the environment is low. As *S. marcescens* begins to grow, the regulator protein

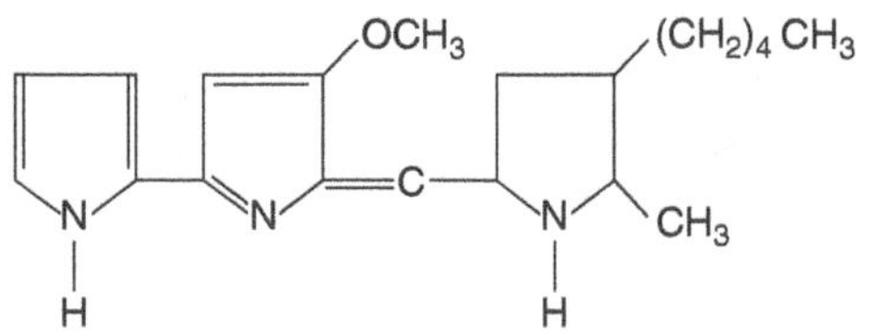

Figure 5.1 Molecular structure of prodigiosin. *Excerpted from Bacterial Synergism Demonstration Kit (Item 154744), © Carolina Biological Supply Company. Used with permission only.*

accumulates in the environment. Once a threshold concentration of the regulator protein is met, all of the organisms in the environment will begin to produce prodigiosin. Therefore, prodigiosin production is cell density dependent (reviewed in Khanafari *et al.*, 2006).

Several laboratories throughout the world are currently studying prodigiosin because it has some interesting properties that can be exploited for the benefit of humankind (reviewed in Khanafari *et al.*, 2006). Studies have implicated prodigiosin as a potential anticancer agent (reviewed in Khanafari *et al.*, 2006). These studies have demonstrated that prodigiosin induces apoptosis (cell death) in several different types of cancer cells while having no effect on normal cells. In addition, it has been demonstrated that prodigiosin is able to inhibit the growth of several different bacteria and fungi (reviewed in Khanafari *et al.*, 2006). Finally, prodigiosin has been implicated as a potential immunosuppressant (reviewed in Khanafari *et al.*, 2006). Immunosuppressants are agents that block the activation of components of the human immune response. They are used, for example, to prevent organ rejection in organ transplant patients. Because of the numerous beneficial properties of prodigiosin described previously, there is a great need to learn as much as we can about the pigment to potentially aid in the development of prodigiosin-based drugs. As such, for the rest of the semester, we will be focusing on the production of this pigment and the biochemistry of this pigment and attempt to determine the environmental conditions required to enhance the production of this pigment by mutants you will generate later in the semester.

Detecting Prodigiosin Spectrophotometrically: Because prodigiosin is a pigment, it can absorb light and can be monitored spectrophotometrically. Haddix and Werner (2000) demonstrated that cell-associated prodigiosin in actively growing cultures maximally absorbs light at 499 nm (Figure 5.2).

For this exercise, you will monitor the production of prodigiosin by *S. marcescens* throughout its life cycle using a technique very

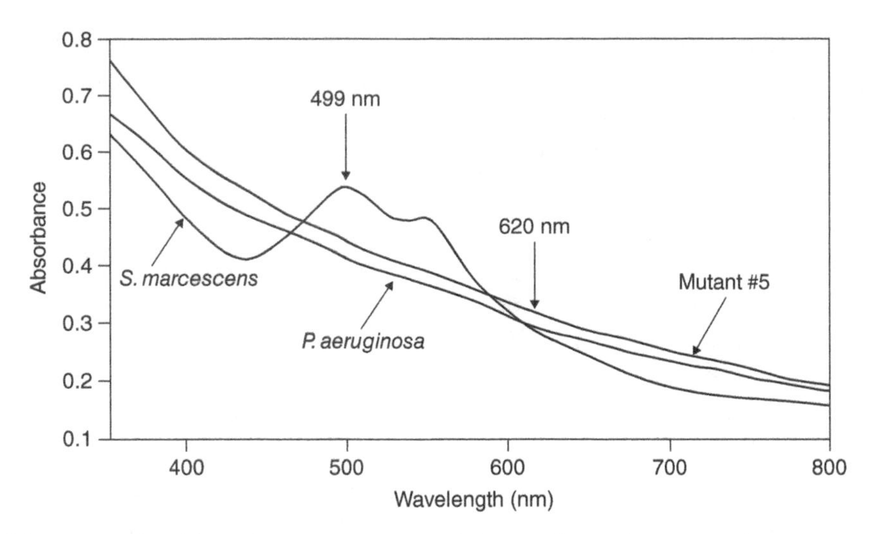

Figure 5.2 Absorbance spectra of bacterial cultures at A_{499}. This figure shows visible spectra from prodigiosin-producing *S. marcescens*, a nonpigmented mutant of *S. marcescens* (mutant 5), and *Pseudomonas aeruginosa* (another bacterium that does not produce prodigiosin) grown under conditions that induce prodigiosin expression (Haddix and Werner, 2000).

similar to the one you used when you worked with *E. coli* to appreciate the bacterial growth curve (Exercise 3). For that exercise, you set the spectrophotometer to measure cell density at a wavelength of 600 nm. Today, you will monitor prodigiosin production by *S. marcescens* at a wavelength of 499 nm and cell density of *S. marcescens* at 600 nm. After you collect your data, you will then use Microsoft Excel to create a graph of your data and perform calculations to determine the units of prodigiosin produced per cell during each phase of the growth curve.

Let's Put This to Practice!

Prior to the start of this laboratory session, stationary-phase cultures of *S. marcescens* were diluted down to create cultures in late-log phase of the bacterial growth curve (optical density = 0.8). You are going to work in groups of 4 to monitor the growth and prodigiosin production by one of these cultures using the spectrophotometer for 3h (180min). You will grow your organisms in a shaking incubator set to 30°C for this study. You will also determine the OD_{600} and A_{499} for a 3-day-old red *S. marcescens* culture. At the end of the 3h period, you will be able to create a graph using Excel to illustrate the amount of prodigiosin produced per stage of the bacterial growth curve.

There will be considerable downtime during this laboratory between each time point. During the downtime, you will be responsible for calculating the amount of prodigiosin produced per cell for each measurement that you take. A discussion of the calculations required follows the protocol. You will also be asked to discuss the Haddix and Werner (2000) paper that you read in preparation for this laboratory exercise:

(1) Turn on the spectrophotometer to allow it to warm up.

(2) Get your workstation ready by clearing the area, washing the lab bench down with 70% ethanol, getting your Bunsen burner ready, and retrieving the supplies/reagents you will need for the lab exercise.

(3) Use sterile technique to remove 2ml of fresh, sterile media into a spectrophotometer cuvette. You will use the fresh media as a blank for the spectrophotometer. The protocol for this is as follows:

(a) Open the media container near the Bunsen burner flame.

(b) Quickly flame the lid of the media container.

(c) Using a sterile pipette, obtain 2ml of the media (be sure not to touch the tip of the pipette to any surfaces excluding the media!).

(d) Pipette the media into the cuvette. Do this in close proximity to the Bunsen burner flame.

(e) Flame the lid of the container.

(f) Close the container lid.

(g) Place the cuvette on the lab bench in close proximity to the Bunsen burner flame.

(4) Use sterile technique to remove 2 ml of your culture into a spectrophotometer cuvette following steps a–g. Once you remove the 2 ml, make sure to return the culture to the 30°C shaking incubator.

(5) Set up the spectrophotometer by ensuring that it will take readings in absorbance units (A) and use the arrow buttons to change the wavelength to 499 nm.

(6) Place your blank cuvette into the spectrophotometer, ensuring that the clear sides of the cuvette are in line with the light path.

(7) Close the lid on the spectrophotometer and hit the 0 button to blank the spectrophotometer. When the machine is blanked, you will see a "0" reading on the view screen. If you do not see a "0," hit the blank button again and wait to see if you get a "0" reading. If you still have trouble, please ask for assistance.

(8) Open the lid and remove your blank.

(9) Place your culture sample into the machine, ensuring that the clear sides of the cuvette are in line with the light path.

(10) Close the lid and record the reading on the view screen. You do not need to press any buttons on the machine to take a reading. The reading is the absorbance units for your culture at 499 nm.

(11) Open the lid and remove your sample.

(12) Change the wavelength on the spectrophotometer to 600 nm.

(13) Place your blank cuvette into the spectrophotometer, ensuring that the clear sides of the cuvette are in line with the light path.

(14) Close the lid on the spectrophotometer and hit the 0 button to blank the spectrophotometer.

(15) Open the lid, remove your blank, and discard it in the red biohazard waste.

(16) Place your culture sample into the machine, ensuring that the clear sides of the cuvette are in line with the light path.

(17) Close the lid and record the reading on the view screen. You do not need to press any buttons on the machine to take a reading. The reading is the optical density for your culture at 600 nm.

(18) Open the lid, remove your sample, and discard it in the red biohazard waste.

(19) Repeat steps 3–18 every half hour for 3 h.

(20) In addition to the last reading at 180 min, measure the OD_{600} and A_{499} values from an old and very red *S. marcescens* culture that was supplied to you by your laboratory instructor (~3 days old). This culture was the one that was used to initially dilute the samples you worked with for this laboratory exercise.

(21) Record the readings in Table 5.1 (180 min; you should end up with 16 readings; 8 at 499 nm and 8 at 600 nm).

(22) Use Excel to create a growth curve of the data for 0–180 min in Table 5.1 (Which columns are important for the creation of your growth curve?).

Calculation of the Production of Prodigiosin Units per Cell: Haddix and Werner (2000) derived a simple calculation based upon their findings to determine the amount of prodigiosin produced per OD_{600} reading. For this laboratory exercise, we modified the calculations proposed by Haddix and Werner (2000) to reflect units of prodigiosin produced per cell. For the modified calculation, you first have to

Table 5.1 Absorbance at 499 nm and optical density at 600 nm of *S. marcescens* over 3 h (180 min).

Time (min)	Absorbance units at 499 nm (AU)	Optical density at 600 nm (OD)
0		
30		
60		
90		
120		
150		
180		
4320 (3 days)		

correct for the impact of the increase in cell density on the absorbance readings you obtained at 499 nm. Despite the fact that cell density changes are maximally detected at an absorbance of 600 nm, there is some interference at 499 nm. To correct for this interference so that the absorbance values you obtained at 499 nm truly reflect prodigiosin production, you must subtract the reading you determined at 600 nm from the reading you determined at 499 nm for each time point. Record your corrected A_{499} values in Table 5.2.

Table 5.2 Correction of A_{499} data for the impact of cell density at 499 nm.

Time (min)	Absorbance units at 499 nm (AU, from Table 5.1)	Corrected absorbance units at 499 nm (AU, A_{499}–OD_{600})
0		
30		
60		
90		
120		
150		
180		
4320 (3 days)		

In Exercise 2, you learned that *S. marcescens* cultures at an $OD_{600} = 1.0$ had 5.8×10^8 cells/ml (Haddix *et al.*, 2000). In order to normalize the units of prodigiosin you detected at each time point to units per cell, you need to multiply this number by your OD_{600} reading at each time point. The resulting value represents the number of cells of *S. marcescens* present per ml at that OD_{600} reading. In addition, you need to account for the sample size read by the spectrophotometer (2 ml). Therefore, to calculate the units of prodigiosin produced per cell, you will need to use the following equation:

Equation 5.1 Calculation to determine units of prodigiosin produced per cell is as follows:

$$\text{Corrected } A_{499} / (OD_{600} \times 5.8 \times 10^8 \text{ cell / ml} \times 2 \text{ ml}) = \text{units of prodigiosin produced per cell}$$

EXERCISE 5B

Record the results from your calculations in Table 5.3.

Finally, the values you determined for Table 5.3 will be very low. In order to make the values more user-friendly, multiply your units of prodigiosin produced per cell value for each time point by 1×10^{10}. Record your results in Table 5.4. Use the information in Table 5.4 to create a graph of the data for 0–180 min to illustrate the production of units of prodigiosin per cell ($\times 10^{-10}$) over time.

Table 5.3 Prodigiosin units produced per *S. marcescens* cell over 3 h (180 min).

Time (min)	Corrected absorbance units at 499 nm (AU, from Table 5.2)	Prodigiosin production (units/cell)
0		
30		
60		
90		
120		
150		
180		
4320 (3 days)		

Table 5.4 Corrected prodigiosin units/cell over 3 h (180 min).

Time (min)	Prodigiosin production (units/cell, from Table 5.4)	Corrected prodigiosin production (units/cell; $\times$ 1×10^{10})
0		
30		
60		
90		
120		
150		
180		
4320 (3 days)		

Reading and Critiquing Primary Scientific Literature: Scientific research papers are the primary means by which scientists communicate significant experimental results. You will be writing up scientific papers describing your experimental results from exercises you performed during this laboratory course. As such, it is necessary for you to be able to read scientific papers with a critical eye to aid you in the development of your own scientific writing skills.

During the past several weeks, you have been discussing and experimenting with a variety of topics related to *S. marcescens* in the laboratory. You have been doing considerable reading in your textbook and laboratory manual on these topics to prepare you for each laboratory session. Your textbook and laboratory manual are considered secondary sources that summarize the scientific results of others and present them to you in a user-friendly format. When you prepared for this laboratory exercise, you had the opportunity to go directly to the primary literature to learn about the experiment you were going to perform. The primary literature reports original research findings as they are discovered. As you may have realized through your reading of the Haddix and Werner paper (2000), reading a scientific paper is not at all like reading a novel or even like reading your textbook. It may be necessary to read and re-read a particular section of a scientific paper before the authors' meaning comes through, but the effort is definitely worth it.

In general, scientists direct their research papers toward the professional audience of a particular journal (just as you will direct your laboratory report toward the faculty reading your reports). Therefore, it may, at first, appear extremely difficult at this stage of your educational career to pick up a scientific paper, read it, and analyze it. However, with practice, the primary scientific literature is quite accessible and will provide to you the most current information available to anyone.

With your laboratory group, discuss the answers to the following questions based upon your reading of the Haddix and Werner paper

(2000). Once each group determines a consensus answer, discuss them as a class:

(1) What was the specific purpose of this study?

(2) What is the hypothesis for this study?

(3) What specific experimental methods were employed to answer the question(s) posed?

(4) What were the controls used for this study.

(5) How were the data presented—graphs, tables, etc.?

(6) Based upon the data obtained, what conclusions did the authors make?

(7) Do you agree or disagree with the author's conclusions? Support your ascertains with facts from the paper.

(8) Is the paper written clearly and organized logically?

Preparation for Laboratory Exercise 6: For the next laboratory session, you will be designing your own experiment to determine experimental conditions to optimize the growth of *S. marcescens* and prodigiosin production. In order to prepare for the laboratory, each group must select an experimental condition to study. Please complete Table 5.5 to indicate which condition your group is interested in studying.

Table 5.5 Environmental conditions that will be studied during Exercise 6.

Experimental condition	Group name
Temperature	
pH	
Carbon source	
Salinity	
Oxygen requirements	

If you own your own laptop, please bring it to lab next session. If you do not, please bring a device to save files.

NAME ______________________________ *DATE* ____________

Exercise 5C: Measuring Prodigiosin Post-laboratory Thinking Questions

Directions

Answer the following questions upon completion of the laboratory exercises.

For the first part of the laboratory, you were asked to use the spectrophotometer to measure the absorbance and optical density of your bacterial culture for 3 h (180 min) at both 499 and 600 nm, respectively. You also took the OD_{600} and A_{499} of an older *S. marcescens* culture. At the end of the laboratory session, you were asked to record the data you obtained in Table 5.1. Please ensure that you have all of the information in Table 5.1 to answer the following questions:

(1) After you completed the absorbance measurements, you were asked to calculate the units of prodigiosin produced per cell. Please write out your calculations for your first time point.

After you completed all of the calculations, you were asked to complete Table 5.4. Please ensure that you have all of the information in Table 5.4 to answer the following questions.

(2) Using the data in Tables 5.1 and 5.4, you were asked to create two graphs using Excel for the 0–180 min time points. The first graph was a growth curve of *S. marcescens* over time (Table 5.1), and the second was to illustrate the production of units of prodigiosin per cell over time (Table 5.4). Please copy/paste the graphs into this worksheet.

(3) Restate the hypothesis you developed for your pre-laboratory thinking questions. Is your hypothesis correct or incorrect? Use the data presented in the graphs of the data from Tables 5.1 and 5.4 and your old, red culture data to support your answer.

(4) How does your data compare to those in the Haddix and Werner (2000) paper? Support your ascertains with facts from your data and the paper.

NAME ______________________________ *DATE* ____________

Exercise 6A: Conditions Affecting the Growth of and Prodigiosin Production by *Serratia marcescens* Pre-laboratory Thinking Questions

Directions

Read over the introduction for this week's laboratory and protocols from last week's laboratory exercises and answer the following questions to ensure that you are prepared for the session:

(1) What are the overall objectives for today's laboratory (provide a numbered list)?

(2) If you have been taking note of what you have been working on for the past several laboratory exercises, you might have noticed some of the conditions you have been using to grow *Serratia marcescens* and measure prodigiosin production. What were they (what type of media did you use, what growth conditions did you use)?

(3) What environmental condition did you agree to test today in the laboratory?

(4) The following supplies and reagents will be available to you in the laboratory today:

(a) Spectrophotometers

The Fundamentals of Scientific Research: An Introductory Laboratory Manual, First Edition. Marcy A. Kelly.

Companion website: www.wiley.com\go\kelly\fundamentals

(b) Spectrophotometer cuvettes

(c) Pipette pumps

(d) Sterile 2 ml pipettes

(e) Micropipettors

(f) Tips for micropipettors

(g) Nutrient broth with 1.0% maltose to use as a blank for the spectrophotometers (tubes will be labeled “blank”)

(h) Late-log cultures of *S. marcescens* D1 labeled as follows:

4°C

30°C

60°C

pH 3

pH 7

pH 11

0.1% NaCl

1.0% NaCl

3.0% NaCl

Aeration (shaking)

No aeration (no shaking)

Nutrient broth

Nutrient broth + 1.0% maltose

Nutrient broth + 1.0% glucose

(i) Shaking incubator set to 30°C

(j) Incubator set to 30°C without shaking

(k) Shaking incubator set to 65°C

(l) 4° refrigerator with rotator for shaking

Using the list of supplies and reagents, design an experiment to test the impact of the environmental condition you selected on the growth of *S. marcescens* and the production of prodigiosin over 3 h

(180 min) (don't forget your controls!). You will need to provide your laboratory instructor with an experimental protocol (which includes controls) prior to the start of your laboratory session. The protocol should be listed in steps, similar to the protocols in this laboratory manual. You will also need to provide your laboratory instructor with a table or figure that you will fill in with data as the laboratory exercise progresses. Your laboratory instructor will review your protocol and may provide comments to assist you prior to the start of the laboratory session. You will not need all of the supplies/reagents listed previously for your experiment.

(5) For your environmental condition you are going to test in the laboratory, hypothesize which of the tests will be optimal with respect to growth for that condition. For example, if you are going to test three different temperatures in the lab today, which of the three different temperatures do you predict will be optimal for the growth of *S. marcescens*? Support your hypothesis.

(6) For your environmental condition you are going to test in the laboratory, hypothesize which of the tests will be optimal with respect to prodigiosin production for that condition. For example, if you are going to test three different pHs in the lab today, which of the three different pHs do you predict will be optimal for prodigiosin production? Support your hypothesis.

NAME ______________________________ *DATE* ____________

Exercise 6B: Conditions Affecting the Growth of and Prodigiosin Production by *S. marcescens*

Introduction

The data you obtained in laboratory Exercises 3 and 5 enabled you to gain an appreciation for the stages of the bacterial growth curve and the production of prodigiosin throughout the life cycle of *S. marcescens*. In this exercise, you will learn more about the optimal conditions required for the organism to grow and produce prodigiosin.

Optimal Conditions for Bacterial Growth: Organisms, in general, require certain environmental conditions in order to grow optimally (at their best). This is because the enzymes that control the metabolism of the organisms do not function properly if the environment is not appropriate for their activity. Specific ranges within the following listed environmental conditions are often evaluated when attempting to determine the conditions required for the optimal growth of most organisms:

Temperature

pH

Oxygen availability

Nutritional sources (especially sources of carbon and nitrogen)

Salinity (salt concentration)

EXERCISE 6B

Humans and a large number of organisms associated with humans function optimally at 37°C. These organisms are considered mesophiles with respect to their optimal temperature requirements. Organisms that function optimally at higher temperatures are considered thermophiles, and organisms that function optimally at lower temperatures are called psychrophiles. With respect to pH, humans and a large number of organisms associated with humans typically function optimally at a neutral pH of 7, categorizing them as neutrophiles. Organisms that prefer acidic conditions are considered acidophiles, and organisms that prefer basic conditions are considered alkaliphiles. Aerobic organisms, such as humans, require oxygen in their environments for growth. Anaerobic organisms, such as many bacteria that live inside humans, do not require oxygen in their environments. Most organisms use glucose as their primary source of carbon and ammonia-containing molecules as their primary sources of nitrogen. With respect to salinity, humans and organisms associated with humans prefer environments with a salinity of approximately 0.9%.

There are many organisms that have vastly different environmental requirements for optimal growth compared to humans and organisms associated with humans. For example, organisms that thrive in the ocean prefer a salinity of approximately 3.5%. These organisms are considered halophiles. Members of domain Archaea typically require extreme conditions such as high heat (extreme thermophiles) or high salinity (extreme halophiles) for optimal growth.

Organisms that can survive at environmental extremes have developed evolutionary adaptations to help them do so. For example, the bacterium, *Helicobacter pylori*, survives in the human stomach. The pH in the human stomach is approximately pH 3. Because *H. pylori* can survive at such a low pH, it is considered an extreme acidophile. Exposure of enzymes to such a low pH would usually denature the enzymes and block the metabolic processes within that organism. The organism would be unable to survive. *H. pylori* is able to circumvent enzyme denaturation by producing enzymes that convert molecules present in the stomach to bicarbonate. Bicarbonate is a buffer. It surrounds the organism and protects it from the pH extreme present in the human stomach. In addition, *H. pylori* is able to degrade stomach mucus to allow it to burrow into the epithelial lining of the stomach. The epithelial lining of the stomach is much less acidic than the lumen of the stomach.

Growth and Prodigiosin Production Optima for *S. marcescens*: Because of the immunosuppressive, anticancer, antibacterial, antifungal, and antiproliferative properties of prodigiosin, several laboratories have set out to discover the optimal conditions for the growth of *S. marcescens* and the production of prodigiosin (reviewed in Khanafari *et al.*, 2006). The intent of these laboratories is to determine a way to maximize the production of prodigiosin so that it might be produced in bulk using the most efficient (and cheapest) method possible. Our ultimate goal is to contribute to this effort as well.

Prodigiosin is produced by a series of enzyme-mediated chemical reactions. You will study the biochemical pathway for prodigiosin production in greater depth in another laboratory exercise. For this week's laboratory, it is important for you to realize that the enzymes required for prodigiosin production function optimally under specific environmental conditions and these conditions may (or may not) be similar to the conditions required for optimal growth.

Today, you will work in groups to determine the optimal temperature, pH, carbon source, salinity, and oxygen requirements for the growth of *S. marcescens* and the production of prodigiosin. Technique-wise, you will be following a protocol that is very similar to the one you used for Exercise 5.

Let's Put This to Practice!

Prior to the start of this laboratory exercise, several stationary-phase cultures of *S. marcescens* were diluted down to 0.5 in the appropriate media. You are going to work in groups of four to monitor the

EXERCISE 6B

Table 6.1 Environmental conditions to be tested during this exercise.

Environmental condition
Temperature (4, 30, 60°C)—these cultures should be grown at the temperatures indicated with shaking
pH (3, 7, 11)—these cultures should be grown at 30°C with shaking
Carbon source (nutrient broth, nutrient broth + 1.0% maltose, nutrient agar + 1.0% glucose)—these cultures should be grown at 30°C with shaking
Salinity (0.1% NaCl, 1% NaCl, 3% NaCl)—these cultures should be grown at 30°C with shaking
Oxygen requirements (shaking at 30°C, no shaking at 30°C)

growth and the production of prodigiosin of these cultures under the environmental condition you selected using the spectrophotometer over 3h (180min). You will use the protocols you developed to carry out these experiments. Upon completion of the laboratory, the class will share the data obtained so that each student can prepare five different Excel graphs based upon the five different conditions tested. The conditions are listed in Table 6.1.

Ensure that you know where all of the supplies and reagents are for this laboratory. Walk around and find the appropriate incubators and materials.

There will be some downtime during this laboratory between each time point. During the downtime, your laboratory instructor will discuss the formal laboratory report assignment with you.

At the end of the session, please share your data with your classmates.

NAME ______________________________ *DATE* ____________

Exercise 6C: Formal Laboratory Report Describing the Conditions Affecting the Growth of and Prodigiosin Production by *S. marcescens*

Introduction

Scientists report the results from their studies in both written and oral format. As a beginning scientist, it is essential that you gain practice in reporting your results using the same format that professional scientists use. During the downtime during the laboratory session, you will also have a chance to discuss the formal laboratory report. Use these discussions and the following criteria to assist you in the preparation of your laboratory report.

General Instructions

Formal laboratory reports must be typed (double spaced, 12-point Times New Roman font, default Word margins) and be between 20 and 25 pages. Tables and figures must be computer generated (not handwritten). You are expected to use proper grammar and correct spelling in your report. Therefore, please be sure to spell and grammar check your document before you submit it. Each section should begin on a new page and be written up as a clear, concise essay, not a list of answers to the points provided in this handout. The listed points should be used as a checklist to guide you through writing the lab report.

You will need to perform web-based research to learn more information than what is presented to you in this lab manual for your report (especially for the introduction section). When performing this research, only use sources from .edu, .org, or .gov websites. DO NOT use Wikipedia as a primary source. You will need to cite AT LEAST 5 sources (other than this laboratory manual) in your report. You may use some of the citations that are listed at the end of this lab manual.

Everything you write in your laboratory report should be in your own words. Summarize any web-based information and then, cite the reference. Do not copy anything verbatim from any source (website, book, a peer, etc.)—using quotations in any form is not acceptable. For your citations, please follow the Council of Science Editors (CSE) citation style. You can most likely find information on this citation format in your college or university library.

Title: Laboratory reports should have a separate title page (include the title and your name). The title for your laboratory report should not be the title used in this laboratory manual; it needs to be more descriptive and include the following information in a single sentence (the information below is not listed in any particular order):

(a) The name of the organism you worked with

(b) The five environmental conditions tested

(c) The two characteristics of the organism that you studied using the five environmental conditions

Abstract: 250 words maximum:

(a) Name of the organism you worked with.

(b) Characteristics of the organism you studied.

(c) Temperatures tested, pHs tested, carbon sources tested, salt concentrations tested, and oxygen requirements tested.

(d) Briefly summarize your results.

Introduction: 3–5 pages. Web-based literature sources should be used to help you write this section:

(a) Background information about *S. marcescens*.

(b) Background information about prodigiosin.

(c) Background information on environmental conditions required for the optimal growth of bacteria.

(d) Discuss how environmental conditions affect the metabolic functioning and, as a result, the growth of bacterial cells.

(e) Discuss environmental conditions required for the growth of *S. marcescens* that you learned about from your own literature search.

(f) Discuss environmental conditions required for optimal prodigiosin production that you learned about from your own literature search.

(g) Describe the objectives for this lab (refer to your pre-laboratory thinking questions).

(h) Briefly summarize what you did to address each objective.

(i) State hypotheses for each of the conditions tested.

Materials and Methods: 1–2 pages. Provide details of what you did in the laboratory in paragraph format.

(a) One paragraph should describe the environmental conditions your lab group tested (with mention of the exact temperatures or pHs that your group tested).

(b) At least 1 paragraph should describe the exact protocol your lab group followed. To do this, you must convert the protocol you wrote with your group for the pre-laboratory thinking questions into a concise paragraph. Be sure that everything is in your own words. The protocol each group member describes here should not be verbatim.

(c) One paragraph should focus on the data you recorded, the calculations you used to convert absorbance units at 499 nm to units of prodigiosin produced per cell, and how you presented the data in this report.

Results: Include the figures and narrative in the order as listed below. Use one figure per page in the report. Be sure to develop your own titles for each of your figures:

(a) Figure 1—Temperature versus growth graph.

(b) Figure 2—pH versus growth graph.

(c) Figure 3—Carbon source versus growth graph.

(d) Figure 4—Salinity versus growth graph.

(e) Figure 5—Oxygen requirements versus growth graph.

(f) Figure 6—Temperature versus prodigiosin production.

(g) Figure 7—pH versus prodigiosin production graph.

(h) Figure 8—Carbon source versus prodigiosin production graph.

(i) Figure 9—Salinity versus prodigiosin production graph.

(j) Figure 10—Oxygen requirements versus prodigiosin production graph.

(k) Briefly summarize the information learned from all of the figures in paragraph format (1–2 pages). Complete this summary in two separate sections—focus the first section on the optimal conditions required for growth and the second section on the optimal conditions for prodigiosin production.

Discussion: 3–5 pages. You will have to do more web-based research and use more citations for this section:

(a) The first paragraph of your discussion should restate the objectives and hypotheses for your study and summarize how you performed the study.

(b) The next several paragraphs should describe your results again. You may use the summaries that you wrote in the results section of your report to assist you with this section, but shy away from directly copying and pasting from your results section. Summarize what you wrote in the results section.

(c) If you ran into any difficulties while you were performing your experiments that may have impacted your results, describe those difficulties here and suggest what impact you think they had on your data.

(d) Do your growth results support the findings by others based upon your literature searches? Why or why not?

(e) Do your prodigiosin production results support the findings by others based upon your literature searches? Why or why not?

(f) Review your data closely. What conditions would you use to enhance prodigiosin production? Keep in mind that the more bacterial present, the more prodigiosin; so, you want to evaluate the growth and prodigiosin production conditions simultaneously to come up with a single set of conditions that could be used to maximally produce prodigiosin.

(g) If you were to continue to work on this project, what would you do next and why? Are there more conditions you would want to test? Why? To answer these questions, you will have to do more web-based research and use more citations.

Citations: Use CSE format (see Appendix A). You should have at least five citations.

NAME ______________________________ *DATE* ____________

Exercise 7A: Biochemistry of Prodigiosin Production Pre-laboratory Thinking Questions

Directions

Read over the introduction and protocols for this laboratory exercise, and answer the following questions to ensure that you are prepared for the session:

(1) What are the objectives for today's laboratory (provide a numbered list)?

(2) What is a biochemical pathway?

(3) What is the relationship between a gene and an enzyme?

(4) What are the amino acids that make up prodigiosin?

(5) Describe the biochemical pathway responsible for the production of prodigiosin.

The Fundamentals of Scientific Research: An Introductory Laboratory Manual, First Edition. Marcy A. Kelly.

Companion website: www.wiley.com\go\kelly\fundamentals

EXERCISE 7B

NAME ______________________________ *DATE* ____________

Exercise 7B: Biochemistry of Prodigiosin Production

Introduction

Biochemical Pathways: Functioning of a cell, such as *Serratia marcescens*, requires active biochemical pathways. Biochemical pathways are responsible for the sequential conversion of reactants into intermediates and, finally, into products by several enzymes. The resulting products can have a structural role in the cell, a functional role in the cell, or could be destined for export from the cell to act elsewhere. Figure 7.1 depicts a hypothetical biochemical pathway.

The enzymes responsible for each individual reaction in the biochemical pathway are specific for that one reaction. They have active sites that interact with a single, specific substrate (whether that is the initial reactant or one of the intermediates of the biochemical pathway) and produce a specific product (whether that is an intermediate of the pathway or the final product of the pathway). The information to synthesize each individual enzyme of a biochemical pathway is encoded in the organisms' DNA. The blueprint for every enzyme in a biochemical pathway is encoded in a single unit of the DNA, called a gene. Therefore, the synthesis of the individual enzymes of a biochemical pathway can be controlled (e.g., activated or repressed) at the genetic level.

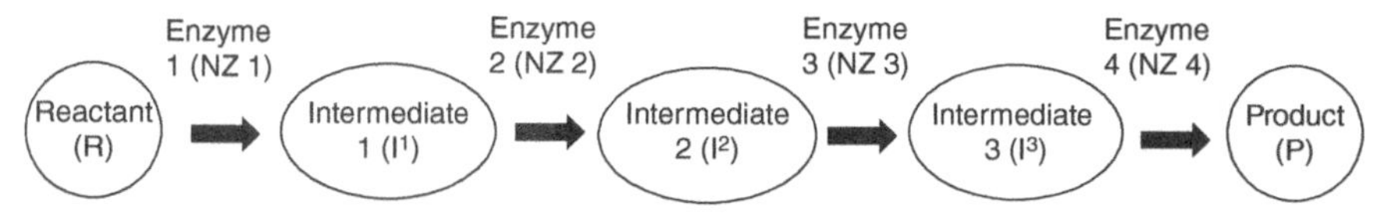

Figure 7.1 A hypothetical biochemical pathway.

Mutations: Because the blueprint for the synthesis of the individual enzymes of a biochemical pathway is in the organisms' genes, any changes to the DNA encoding the enzymes can impact the control or the function of the enzymes. Changes in a DNA sequence results in a mutation. Mutations in a gene encoding an enzyme involved in a biochemical pathway may render the gene, and, as a result, the enzyme, nonfunctional. Wild-type organisms are organisms that have fully functional biochemical pathways with no mutations in the genes encoding the enzymes of their biochemical pathways. Auxotrophs are organisms that have mutations in one or more of the genes encoding enzymes of a biochemical pathway. These organisms are unable to convert reactants to products in that biochemical pathway because the enzyme(s) of the pathway are not functional.

Figure 7.2 depicts a sample biochemical pathway with four enzymes for a wild-type organism and four auxotrophic organisms. Each of

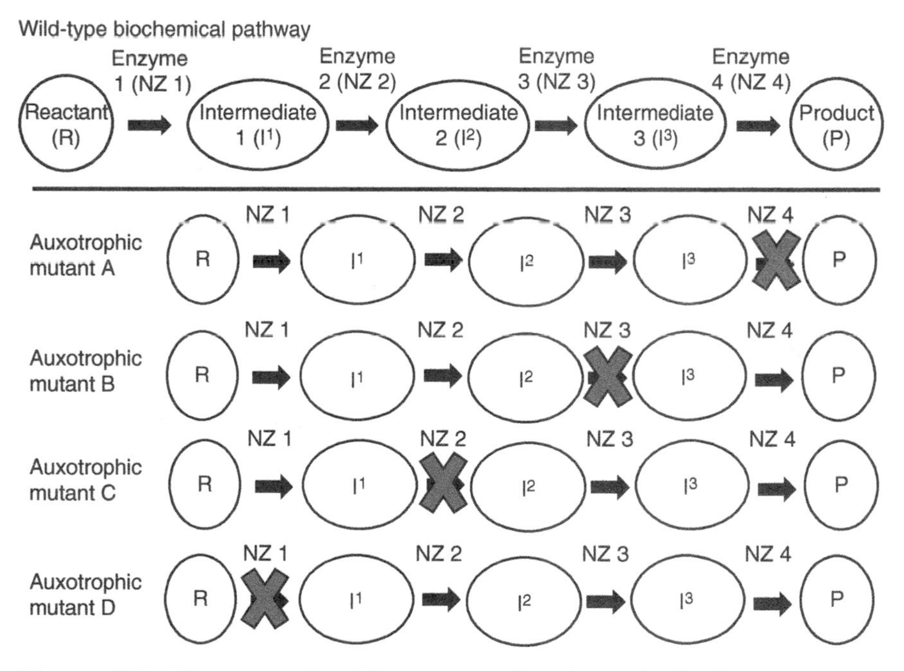

Figure 7.2 Four auxotrophic mutants in a hypothetical biochemical. The nonfunctional enzyme in each mutant is indicated by a red "X". (*See insert for color representation of the figure.*)

the four auxotrophic organisms has a mutation in a different gene encoding enzymes in the biochemical pathway—rendering that enzyme nonfunctional. Many biochemical pathways are controlled by a phenomenon called feedback inhibition. In feedback inhibition, accumulation of the final product of a biochemical pathway turns off the biochemical pathway. Once the final product is depleted, the pathway turns back on to synthesize more product until there is enough to activate feedback inhibition again. Auxotrophic mutants are defective in feedback inhibition because they are unable to synthesize product. Therefore, the biochemical pathway is never turned off and intermediates before the nonfunctional enzyme accumulate.

Deduction of Biochemical Pathways: Scientists use auxotrophic mutants to help them deduce the enzymes and steps involved in biochemical pathways. They can do this because the intermediates produced just prior to the nonfunctional enzyme in a biochemical pathway build up. Review Figure 7.2 again, and for each auxotroph, determine which intermediate accumulates and record your answers in Table 7.1.

Look at Figure 7.2 again but from a different perspective. Determine which enzymes are functional in each auxotrophic pathway and record your answers in Table 7.2.

Compare the information you recorded in Tables 7.1 and 7.2. You should notice that, for example, auxotrophic strains B, C, and D are

Table 7.1 Accumulated intermediates by auxotrophs with blockages at different stages in a biochemical pathway.

Mutant	Accumulated intermediate
A	
B	
C	
D	

Table 7.2 Functional and nonfunctional enzymes in each auxotrophic mutant.

Mutant	Inactive (nonfunctional) enzyme	Active enzymes
A		
B		
C		
D		

Table 7.3 Intermediates have the ability to feed the biochemical pathways of the different auxotrophic mutants.

Mutant	Accumulates…	And would allow the production of product by strains…
A	I3	B, C, and D
B		
C		
D		

all normal for enzyme 4 and would be able to convert intermediate 3 to product if it were supplied to them. Based upon the information in Table 7.1, which of the four auxotrophic strains accumulates intermediate 3? _____

To continue this analysis, use the information in Tables 7.1 and 7.2 to determine which intermediates could be used to complete each of the biochemical pathways for the other three auxotrophic strains. Use Table 7.3 to record your answers.

Hopefully, this analysis helped you to realize that auxotrophic strains with nonfunctional enzymes later in a biochemical pathway can "feed" strains with nonfunctional enzymes that exist earlier in that same pathway to produce product. This information can be used to dissect biochemical pathways and determine the sequence of enzymes in that pathway. Specifically, if each strain is grown together, one can determine whether or not they feed each other

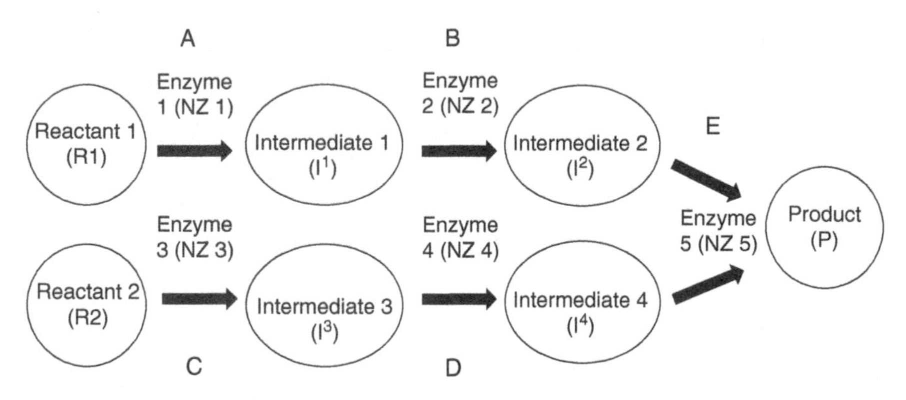

Figure 7.3 A hypothetical branched biochemical pathway.

and, then, deduce the pathway in a similar fashion to what you did for this exercise.

Branched Biochemical Pathways: Not all biochemical pathways are linear—many are branched. An example of a branched pathway is depicted in Figure 7.3. For the branched biochemical pathway in Figure 7.3, two separate reactants are converted into intermediates that are joined together to form a final product.

In Figure 7.3, different auxotrophic strains with different blockages are labeled A–E. In dissecting the feeding for branched pathways, similar rules apply as described previously, but in addition, if one branch of a pathway is nonfunctional and one is functional, the two strains can mutually feed each other. Complete Table 7.4 incorporating what you learned from the first part of this exercise and this new rule to assist you.

Prodigiosin Synthesis in S. marcescens: Prodigiosin is composed of several different amino acids including proline, histidine, methionine, and alanine. It is synthesized by a branched pathway. The pathway is depicted in Figure 7.3.

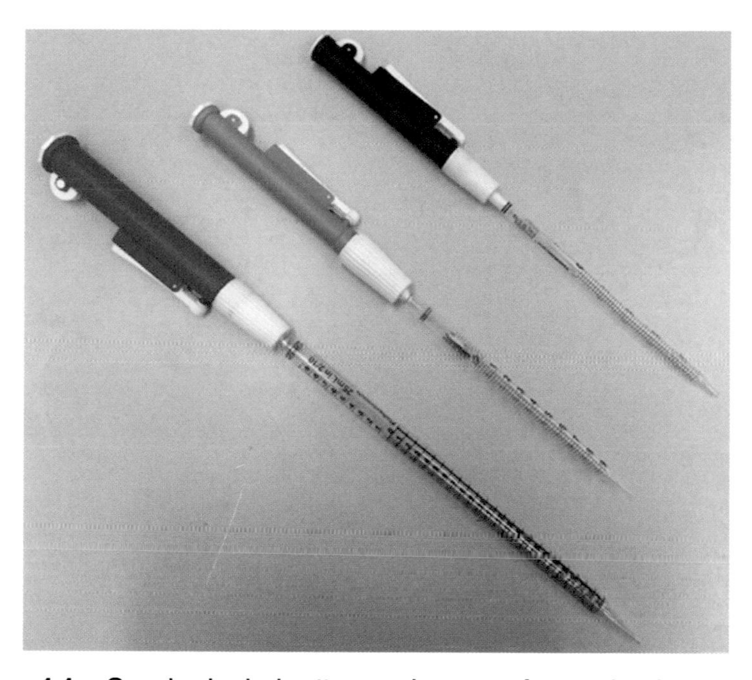

Figure 1.1 Serological pipettes and pumps for total volumes of 25 ml (red pump), 10 ml (green pump), and 5 ml (blue pump).

The Fundamentals of Scientific Research: An Introductory Laboratory Manual, First Edition. Marcy A. Kelly.

Companion website: www.wiley.com\go\kelly\fundamentals

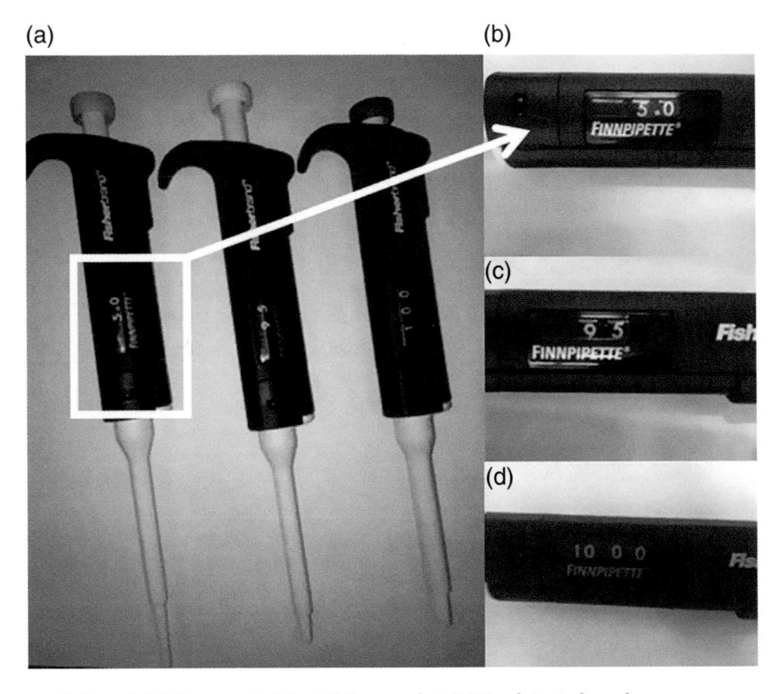

Figure 1.2 (a) Typical 10, 100, and 1000 ul total volume micropipettes, respectively. (b) 10 ul micropipette volume indicator. The volume indicator on a 10 ul micropipette is read from left to right. Digits to the left of the decimal point indicate uls, and digits to the right of the decimal point indicate tenths of uls. (c) 100 ul micropipette volume indicator. The volume indicator on a 100 ul micropipette is also read from left to right. Digits indicate uls up to 100 ul. (d) 1000 ul micropipette volume indicator. The volume indicator on a 1000 ul micropipette is read from left to right. Digits indicate uls up to 1000 ul.

(a)

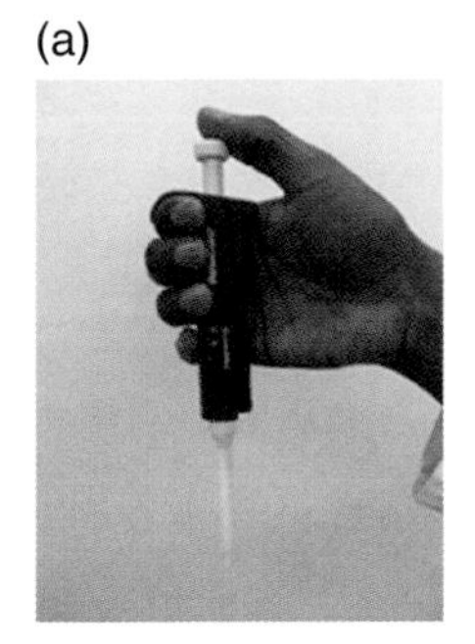

(b)

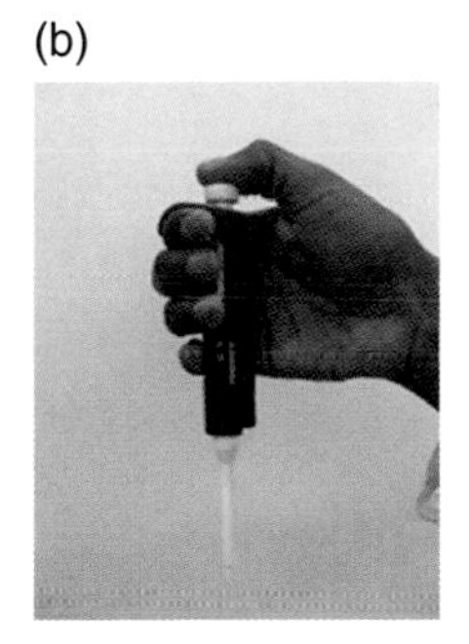

(c)

Figure 1.3 (a) Up position of the plunger button on a micropipette. This is the starting position for proper pipetting. (b) The plunger button is depressed to the first stop position to initiate pipetting. While holding the plunger button in this position, insert the micropipette tip into the sample and then slowly release the plunger button to the up position. The desired volume of sample will be drawn up into the tip. (c) When the plunger button is depressed beyond the first stop position to the second stop position, the liquid in the pipettor tip will be completely expelled.

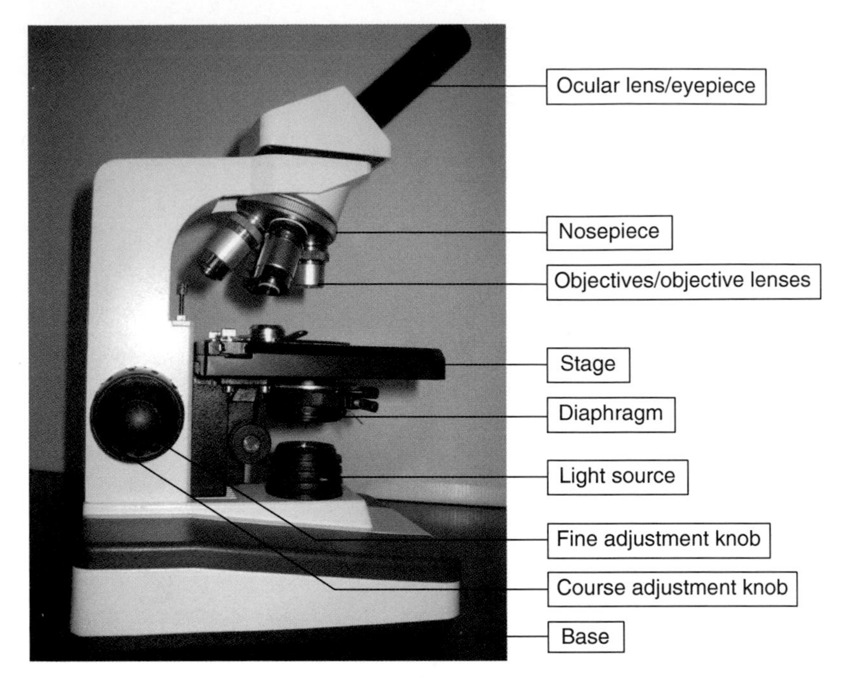

Figure 2.2 Components of the compound light microscope.

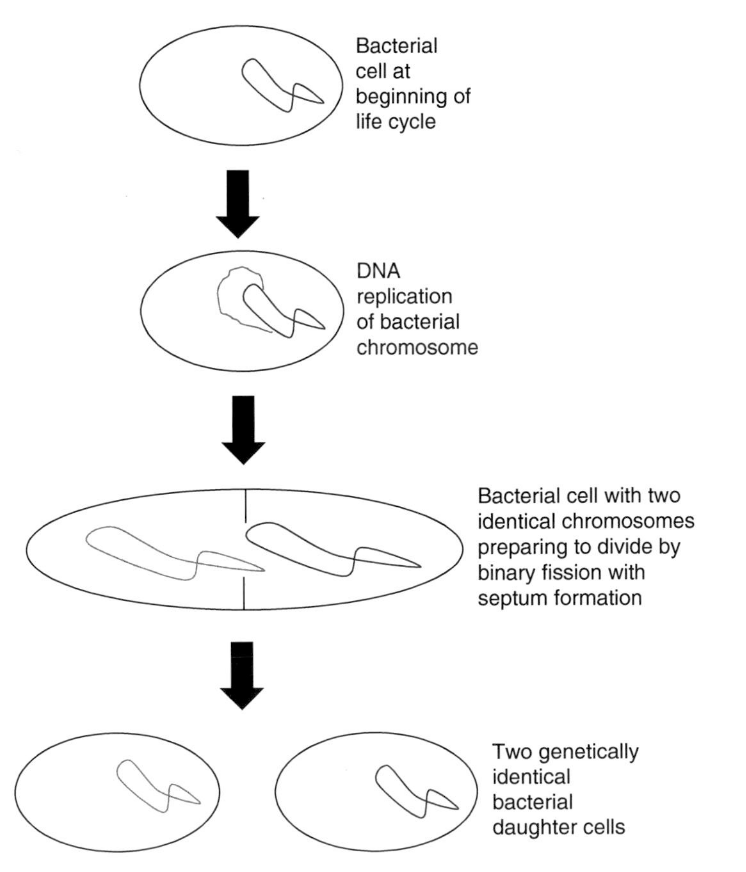

Figure 3.1 Bacterial cell cycle; binary fission.

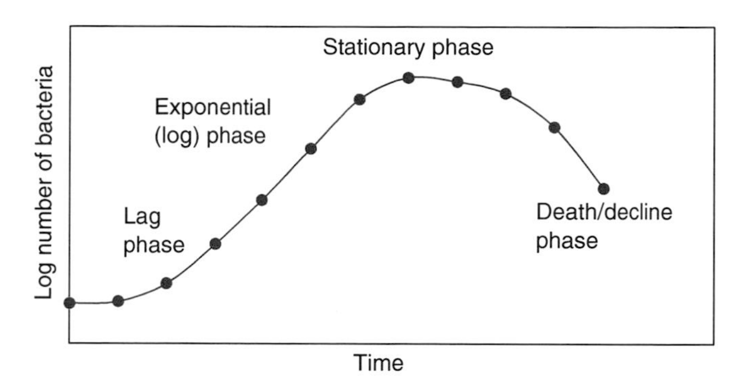

Figure 3.2 The bacterial growth curve.

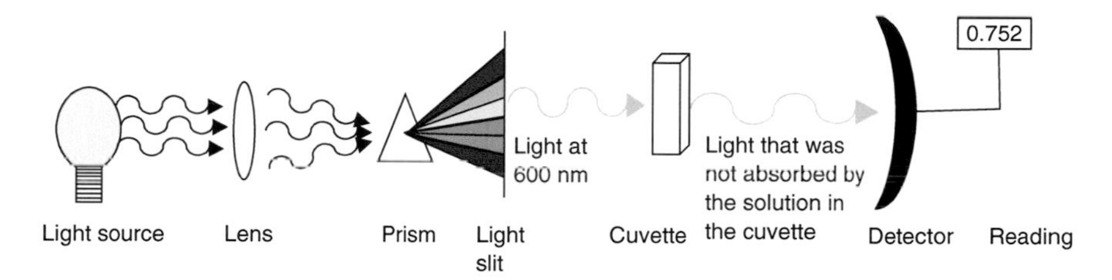

Figure 3.3 The functional components of a spectrophotometer.

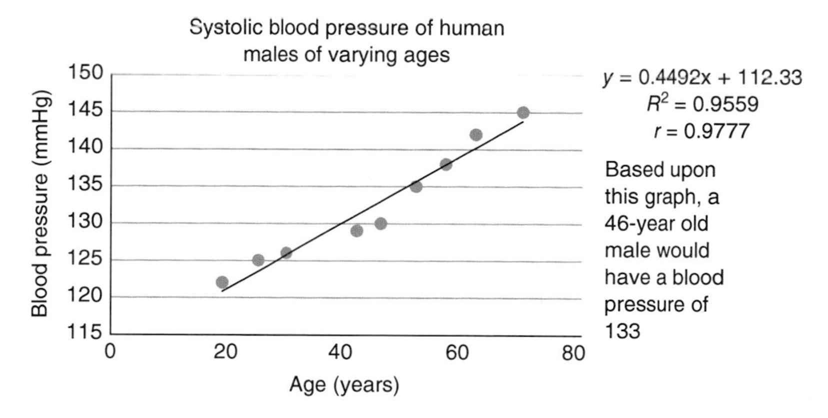

Figure 3.4 Blood pressure Excel sample problem graph.

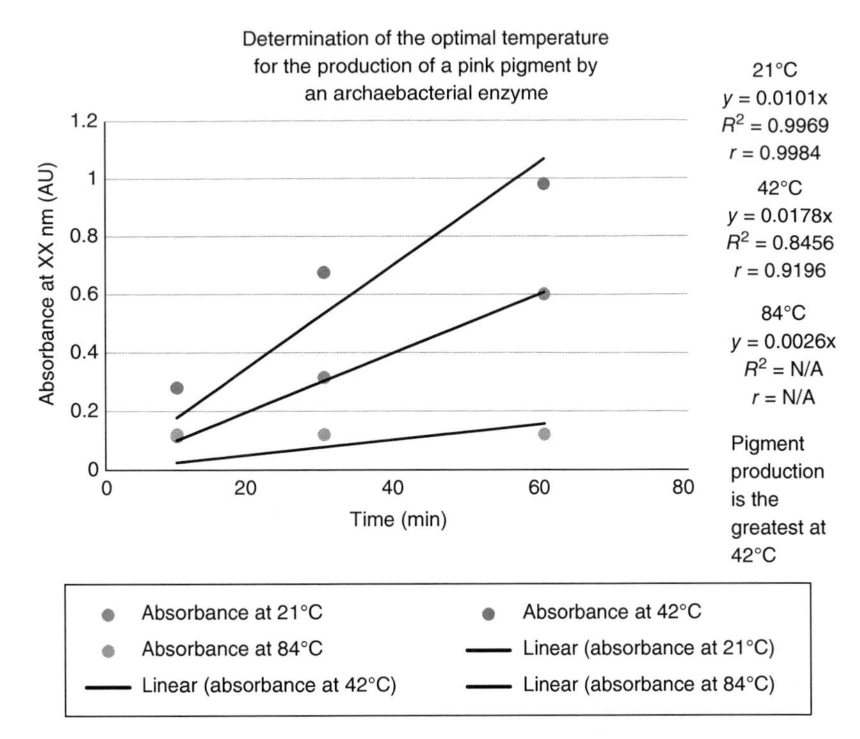

Figure 3.5 Archaeal pigment Excel sample problem graph.

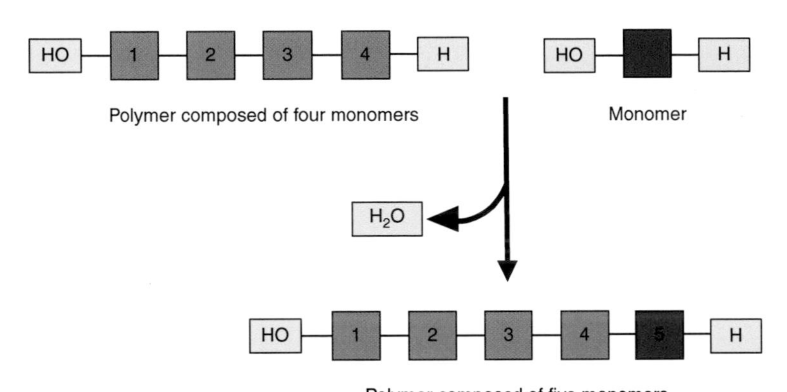

Figure 4.1 Dehydration synthesis between a polymer of four monomers and a single monomer to yield a polymer of five monomers. A single molecule of water is lost during the reaction.

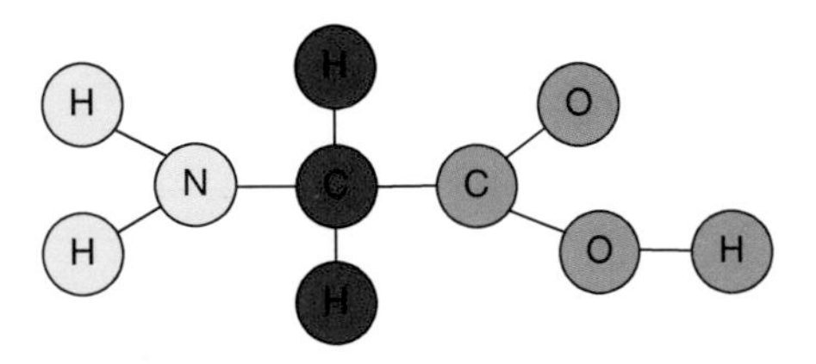

Figure 4.2 An amino acid. Illustration includes the amino group (yellow), central or alpha carbon, and associated R-group which is variable in each amino acid (red), and the carboxyl group (blue).

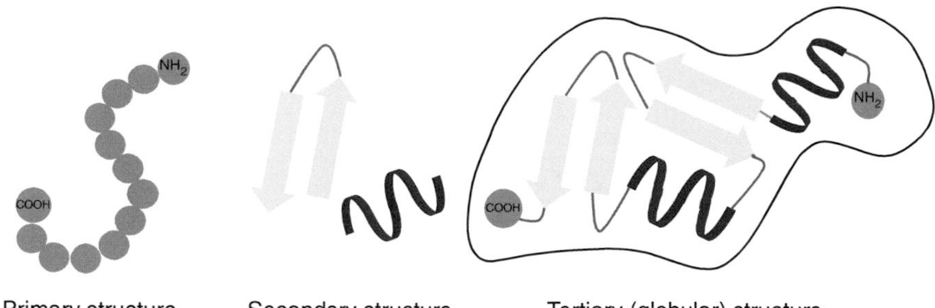

Figure 4.3 Protein folding. The primary structure of a protein consists of a linear chain of amino acids (blue circles). The secondary structure consists of beta pleated sheets (yellow arrows) and alpha helices (red spirals). The tertiary structure is a globular functional protein consisting of secondary structures folded in a specific, coordinated fashion to ensure that protein structure enables protein functioning.

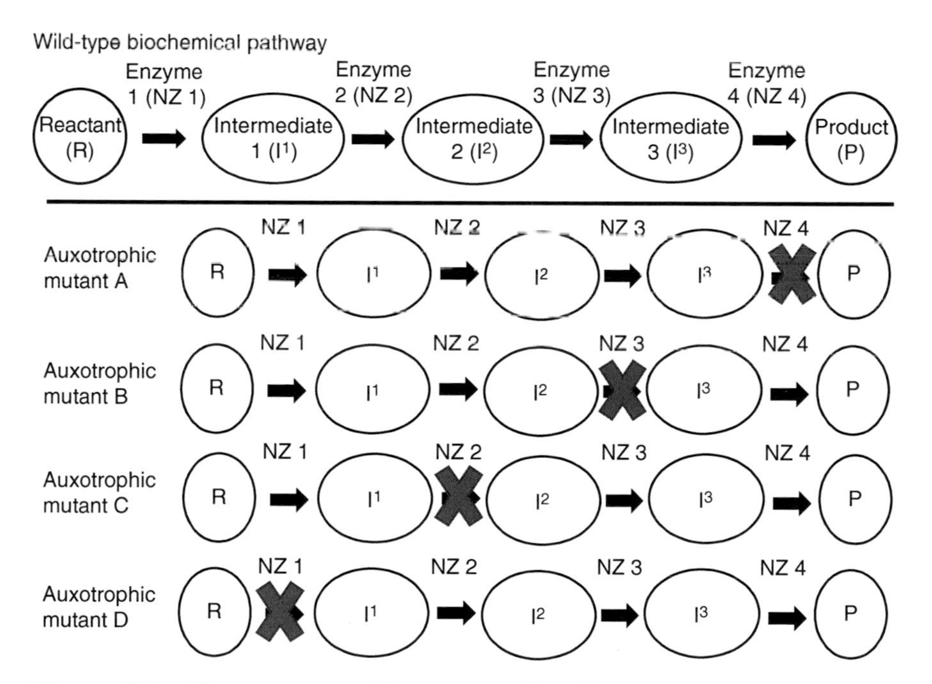

Figure 7.2 Four auxotrophic mutants in a hypothetical biochemical. The nonfunctional enzyme in each mutant is indicated by a red "X".

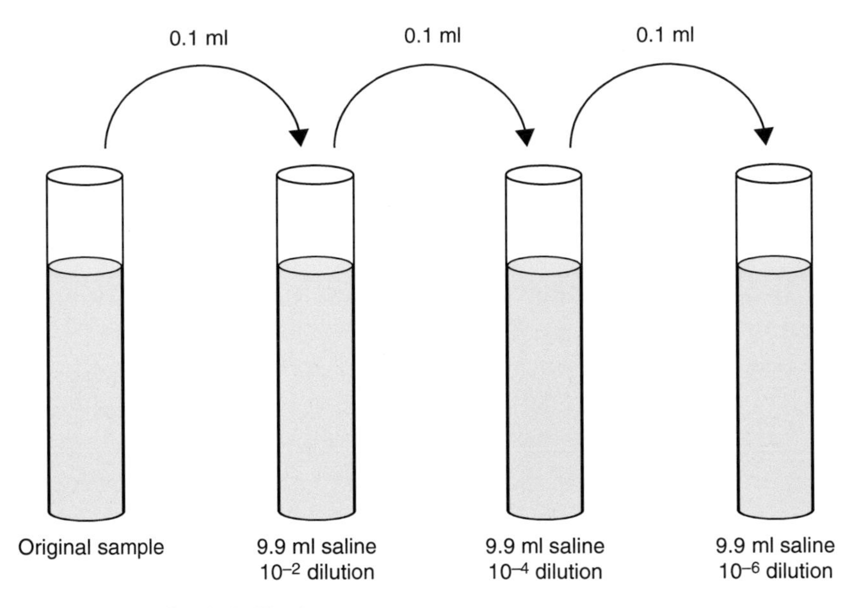

Figure 9.1 Serial dilutions.

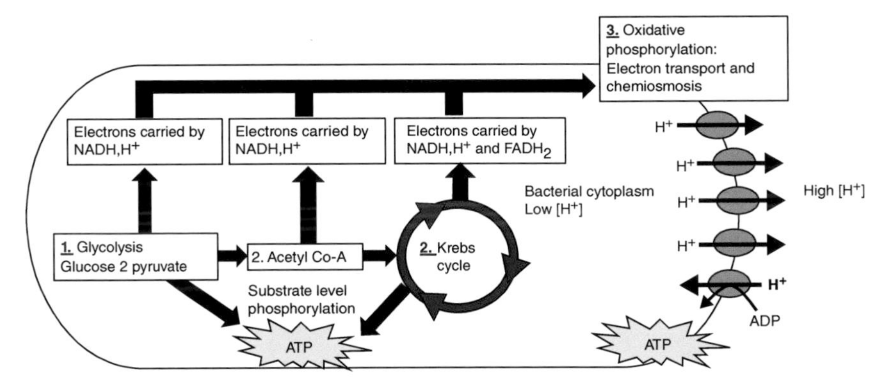

Figure 10.1 Bacterial cellular respiration.

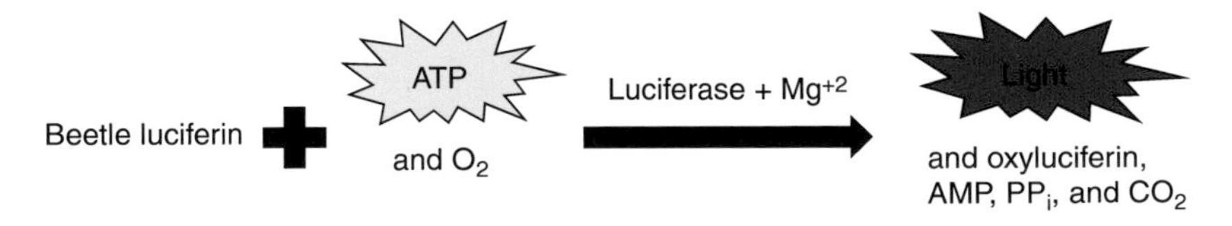

Figure 10.2 The luciferase reaction to quantify the amount of ATP produced by wild-type and mutant *S. marcescens*.

Figure 1. Stationary phase, red *S. marcescens* grown in Nutrient Broth plus 1.0% maltose.

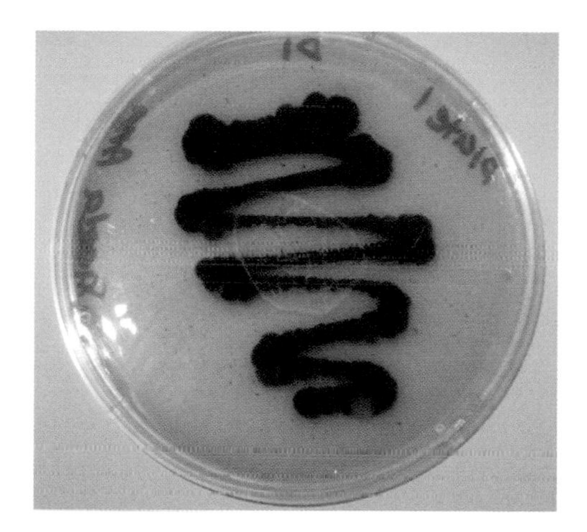

Figure 2. *Serratia marcescens* D1 grown in Nutrient Agar plus 1.0% maltose after 48 hours at room temperature.

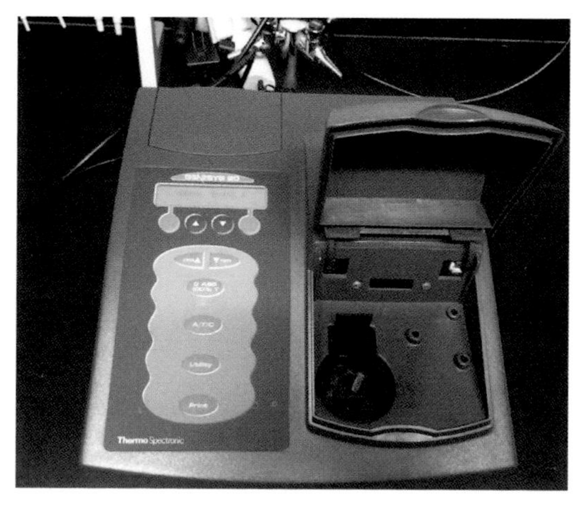

Figure 3. Visible light spectrophotometer.

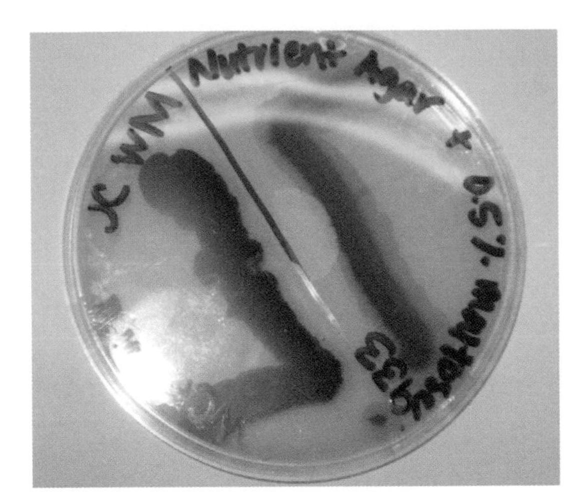

Figure 4. *Serratia marcescens* 933 and WCF prodigiosin null mutants "feeding" each other to mutually form prodigiosin.

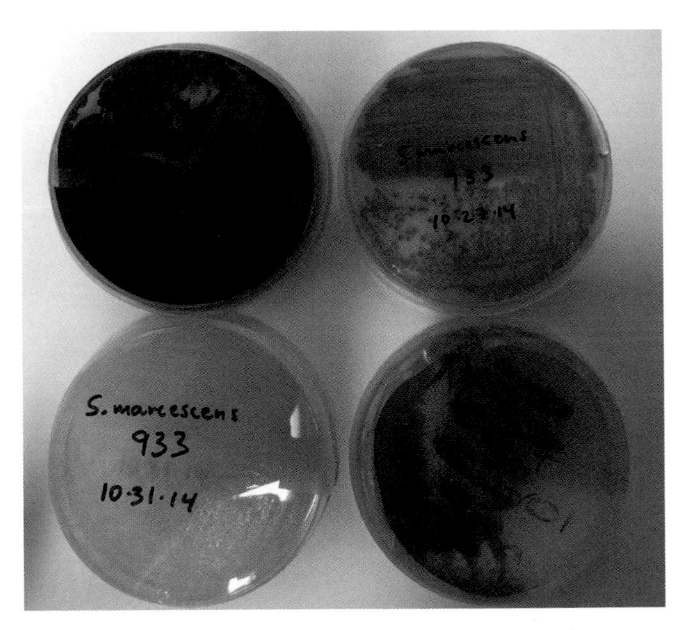

Figure 5. Variations in coloration that may be detected on agar plates inoculated with *Serratia marcescens* 933. The top left plate depicts the prodigiosin produced by *S. marcescens* D1 to serve as a control. The bottom left plate depicts the *S. marcescens* 933 null mutant lacking a pigment. The two plates on the right depict alternative pigments that *S. marcescens* D1 may produce. These pigments are not prodigiosin.

Table 7.4 Intermediates have the ability to feed the biochemical pathways of different auxotrophic mutants in a branched biochemical pathway.

Mutant	Accumulates...	And would allow the production of product by strains...
A		
B		
C		
D		
E		

In today's laboratory, you will study two of the five auxotrophic mutants of *S. marcescens* depicted previously. These auxotrophic strains were produced by exposure of the wild-type strain of *S. marcescens* to ultraviolet light. To study these strains, you will perform pairwise feeding trials to determine which two mutants you have. Although the theory behind how to dissect the sequence of enzymes involved in a biochemical pathway is complicated, the actual procedure for the experiment to do this is quite simple. The strains you will be working with behave similarly compared to wild-type *S. marcescens* with respect to growth. The only difference between the mutants and the wild type is that the mutants, if grown alone, will appear white in color (not red). If an auxotrophic strain turns red while in the presence of another strain, feeding has occurred. In this exercise, you will inoculate the three strains (wild type and the two mutants) in pairs onto plates and evaluate their coloration to help you determine which mutants you are working with.

Let's Put This to Practice!

(1) Get your workstation ready by clearing the area, washing the lab bench down with 70% ethanol, getting your Bunsen burner ready, and retrieving the supplies/reagents you will need for the lab exercise.

(2) Obtain and label 3 agar plates as follows:

(a) Wild-type control

(b) Auxotrophic mutant 1 control

(c) Auxotrophic mutant 2 control

(3) Obtain three additional agar plates and draw a line down the center of the back of each plate to split them into two sections. Label each section with the following names:

(a) Plate 1—wild type and mutant 1

(b) Plate 2—wild type and mutant 2

(c) Plate 3—mutant 1 and mutant 2

(4) Using sterile technique, obtain a loopful of bacteria from the wild-type broth culture that was supplied to you and inoculate the wild-type control plate (2a) as depicted in Figure 7.4. Your instructor will demonstrate how to do this.

(5) Inoculate the mutant 1 (2b) and mutant 2 (2c) control plates in the same manner.

(6) Using sterile technique, obtain a loopful of bacteria from the wild-type broth culture that was supplied to you, and inoculate position 1 on the wild-type/mutant 1 plate as depicted in Figure 7.5

Figure 7.4 Inoculation of control plates.

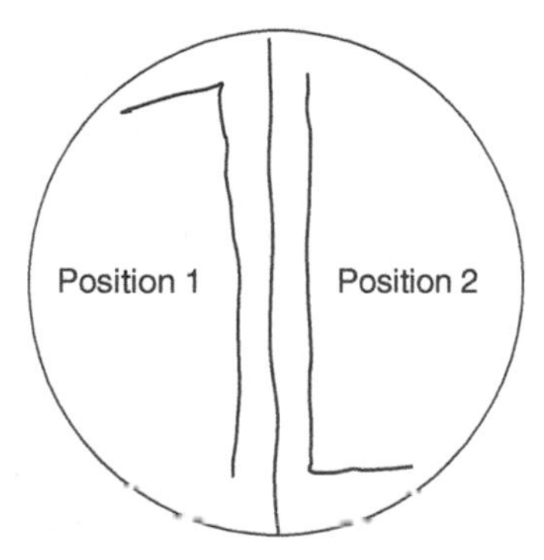

Figure 7.5 Inoculation of test plates.

(upside down "L"). You label position 1 with the name "wild type" on your plate.

(7) Using sterile technique, obtain a loopful of bacteria from the mutant 1 broth culture that was supplied to you and inoculate position 2 on the wild-type + mutant 1 plate as depicted in Figure 7.5 (right side up "L"). Make sure that you leave approximately 1 mm between the two lines of your Ls. You label position 2 with the "mutant 1" on your plate.

(8) Repeat steps 6 and 7 for the wild-type/mutant 2 and mutant 1/ mutant 2 plates, making sure you use sterile technique and the correct cultures for each inoculation.

(9) Seal each inoculated dish with parafilm (your instructor will show you how to do this) to prevent entry of air into the plate and drying out of the agar.

(10) Incubate each dish right side up (agar side down) at room temperature.

(11) In the lab the following week, carefully examine the growth from your three control and three "feeding" trials to determine which auxotrophs show the formation of prodigiosin and which do not. If you detect any red color (regardless of whether the bacteria are uniformly red or red on the edges of growth only), record that as positive prodigiosin production. Record your results in Table 7.5.

After recording your results, discuss them with your group and refer to Figure 7.3 and Table 7.4 to help you determine which two auxotrophic mutants you worked with for this laboratory. Which two mutants did you work with? _____ and _____

Table 7.5 Results of feeding experiments.

Position 1	Position 2	Color of bacterial growth
Wild-type control	NA	
Mutant 1 control	NA	
Mutant 2 control	NA	
Wild type	Mutant 1	
Wild type	Mutant 2	
Mutant 1	Mutant 2	

Once you complete your discussion with your group, your instructor will discuss your results as a class. You may wish to review the information in Appendix B on your own prior to the discussion.

If you own your calculator, please bring it to the lab next session. If you do not, please let your instructor know.

NAME ______________________ *DATE* ____________

Exercise 7C: Biochemistry of Prodigiosin Production Post-laboratory Thinking Questions

Directions

Answer the following questions upon completion of the laboratory exercises.

For the laboratory, you were asked to perform feeding experiments using different strains of *S. marcescens*. You were then asked to record the data you obtained in Table 7.5. Please ensure that you have all of the information in Table 7.5 to answer the following questions:

(1) After you recorded your data, you were supposed to work with your group to determine which two stains you had inoculated onto your plates. What two strains were you working with? How did you come to this conclusion?

(2) Review the information in Appendix B following the analysis of your data. Support the claims you made for question 1 using the biochemical information that was provided to you in the appendix. Use the specific names of the molecules in your discussion.

(3) Refer to Figure 7.3 in this exercise. Propose what would happen with respect to prodigiosin production if you were to grow strains (a) A and E together, (b) C and E together, and (c) A and B together. Support your answers using details from Appendix B.

NAME ______________________________ *DATE* ____________

Exercise 8A: The Probability Basis for Mutation Rate Calculation: A Dice-Roll Exercise Pre-laboratory Thinking Questions

Contributed by Pryce L. Haddix, PhD

(1) Define the following terms or concepts in your own words:

(a) Mutation

(b) Spontaneous mutation

(c) Probability

(d) Random

(2) What is the theoretical (not measured or calculated) probability of achieving a "12" result by rolling a 20-sided die?

(3) Consider the growth of three bacterial cultures. Each of the cultures grew from a single, nonmutant founder cell to one million cells, but they produced different numbers of mutants. Culture 1 produced no mutants; Culture 2 produced 27 mutants, and Culture 3 produced 385 mutants. How may this discrepancy, or fluctuation, in the number of recovered mutants be explained if the mutation rate is assumed to be a constant value of 1 mutation per 500,000 cell division events?

The Fundamentals of Scientific Research: An Introductory Laboratory Manual, First Edition. Marcy A. Kelly.

Companion website: www.wiley.com\go\kelly\fundamentals

NAME ______________________________ *DATE* ____________

Exercise 8B: The Probability Basis for Mutation Rate Calculation: A Dice-Roll Exercise

Contributed by Pryce L. Haddix, PhD

Please remember to complete the work required for Exercise 7 prior to starting this laboratory.

Introduction

Mutations: In Exercise 7, you used auxotrophic mutants of *Serratia marcescens* that were unable to produce prodigiosin to appreciate the biochemistry of prodigiosin production. Each mutant had a single mutation in it that inactivated one of the enzymes in the biochemical pathway to produce prodigiosin. As described in Exercise 7, mutations are changes in the DNA sequence. These changes can be passed down to each subsequent generation of that organism—as long as the mutations do not cause a lethal effect. Mutations are primarily responsible for the genetic diversity seen in organisms.

Mutations can be induced (or caused) by exposure of the organism to mutation-causing chemical or physical agents (aka mutagens) or they can occur spontaneously. We will study the impact of mutagens on prodigiosin production in *S. marcescens* in Exercise 9.

Spontaneous mutations are changes to the genome which occur during normal cellular function in the absence of a known cause. Spontaneous mutations may result from errors by the machinery responsible for copying DNA during DNA replication. The machinery "accidentally" incorporates an incorrect base into the new DNA that it is making. This wrong base is then inherited by the offspring of that cell. In a bacterial culture derived from a single, non-mutant organism, the earlier a spontaneous mutation occurs in that culture, the more offspring that mutation will acquire. The spontaneous mutation rate for *E. coli* has been calculated as roughly one altered base per 10^{10} bases. You may note that this number is extremely small, and as such, the occurrences of spontaneous mutations are rare events.

Probability: Probability is the branch of mathematics which measures the frequency of particular results when those results do not always occur (such as spontaneous mutations). The probability of success is defined simply as the measured number of successes over a large number of events and is usually expressed as a fraction or percentage:

Equation 8.1

Calculation to determine the probability of the success of an event.

Total number of successes/total number of events producing both successes and failures

If the probability for success is low, a large number of events is necessary for accurate measurement of that probability. For example, consider a six-sided die. One roll of that die will likely not produce a "six" result. However, on average, 600 rolls will produce very nearly one hundred "six" results and a probability or *P* value of 1/6 or 17%.

The probability calculation developed by Poisson, known as the Poisson distribution, describes the occurrence of a success over a very large number of events when the probability for success at each event is extremely low. This mathematical model was first applied by Luria and Delbrück to the measurement of spontaneous mutation in *E. coli* (Luria and Delbrück, 1943). These are the investigators that showed that spontaneous mutation of a given gene occurs both at random and at very low frequency. This contribution earned Luria and Delbrück a Nobel Prize in 1969.

When applying the Poisson equation to a growing bacterial cell population, Luria and Delbrück considered an "event" to be a cell division and a "success" to be the production of a mutant cell as the population multiplied from a single, nonmutant founder cell. Their equation may be expressed as

Equation 8.2

Application of the Poisson equation to a growing bacterial cell population.

$$\mu = -(\ln[(P(0)])/N$$

where μ is the probability of mutation, or mutation rate, expressed as the number of mutants per cell division; $P(0)$ is the measured probability of achieving no mutants after N cell division events—this is determined experimentally as the fraction of multiple, identical cultures which produces no mutants after each culture has undergone N cell division events from a single, nonmutant founder cell; and N is the number of cell division events, which is the same for all cultures. N is essentially identical to the measured number of cells in the grown population.

Let's Put This to Practice!

In order to illustrate the use of the equation described previously, we will apply an equivalent version of it to calculate the probability of achieving a "12" result from rolls of a 20-sided die:

(1) Each student should retrieve a 20-sided die from your instructor.

Table 8.1 Number of "non-12" and "12" results upon throwing a die 20 times.

	Hashtags representing number of times a "non-12" number or "12" number is rolled
Non-12	
12	

(2) Roll the die 20 times. As you do, record your individual data both as the number of "non-12" results (numbers 1–11 and 13–20) and as the number of "12" results by using hashtags for each "non-12" in the indicated box in Table 8.1 and hashtags for each "12" result in the indicated box in Table 8.1.

(3) Collect class data and record it in Table 8.2 by individual student.

(4) Once the class data is obtained, each student should complete the calculations required to complete the table.

(5) Using the data in the table and the equation below, determine the measured probability (P) of a "12" roll.

P of a "12" roll = (class total number of "12" rolls/class total number of rolls) = _____.

Analysis of Classwide Data for Calculated Probability Determination: To do this, we will need to adapt the mutation rate equation described in the introduction of this exercise to calculate the probability of achieving a "12" result. Since the mutation rate is conceptually equivalent to the probability we want to calculate in

Table 8.2 Class data for the number of "non-12" and "12" results.

Name	Number of "12" rolls	Number of "non-12" rolls	Total number of rolls
Example	2	18	20
Class totals			

the dice-roll exercise, we may express the mutation rate equation in terms more appropriate to our application:

Equation 8.3

Mutation rate equation.

$$P_{12} = -\left(\ln\left[P_{12}(0)\right]\right) / n$$

For this exercise, $P_{12}(0)$ was measured experimentally as the fraction of students in the class who did not achieve a "12" result after 20 rolls. Determine this result from Table 8.2 and record that information below. Similarly, n was 20, since each student rolled his or her die 20 times:

Number of students who did not achieve a "12" result: _____.

Total number of students: _____.

$P_{12}(0)$ = students with no "12" result/total number of students = _____.

ln ($P_{12}(0)$) = _____; use this value in the equation earlier to calculate the calculated probability (P_{12}) of obtaining a "12" roll.

Calculated probability, P_{12} = _____.

NAME ______________________________ *DATE* ____________

Exercise 8C: The Probability Basis for Mutation Rate Calculation: A Dice-Roll Exercise Post-laboratory Thinking Questions

Contributed by Pryce L. Haddix, PhD

(1) How does the calculated value for rolling a "12" on a 20-sided die compare with the value measured from classwide data? If there is a large discrepancy, what does that mean?

(2) If more students participated in this exercise, what would be the effect on the measured probability?

(3) On the calculated probability?

(4) If each student in your class were to roll the die 100 times instead of 20 times, how would that affect the measured and calculated probabilities?

(5) Imagine that 20 additional students participated in this activity. They each rolled their die 20 times and recorded their "12" and "non-12" results in a similar fashion as what you did for this exercise. How would this modification of the dice-roll activity be incorporated into a mutation rate determination experiment?

NAME ______________________________ *DATE* ____________

Exercise 9A: Understanding Evolution by the Generation of UV Light-Induced Prodigiosin Mutants Pre-laboratory Thinking Questions

Directions

Read over the introduction and protocols for this laboratory exercise and answer the following questions to ensure that you are prepared for the session:

(1) What are the objectives for today's laboratory (provide a numbered list)?

(2) Discuss an example of a selective pressure that has led to the evolution of a species.

(3) What are serial dilutions and what are they used for?

(4) What selective pressure will you subject *Serratia marcescens* to in this laboratory exercise? At the molecular level, what type of mutation does this pressure produce?

(5) After you expose your bacteria to the selective pressure, you are asked to limit exposure to visible light. Why?

(6) As you increase exposure time to the selective pressure, what do you hypothesize will happen to the *S. marcescens*?

The Fundamentals of Scientific Research: An Introductory Laboratory Manual, First Edition. Marcy A. Kelly.

Companion website: www.wiley.com\go\kelly\fundamentals

NAME ______________________________ *DATE* ____________

Exercise 9B: Understanding Evolution by the Generation of UV Light-Induced Prodigiosin Mutants

Introduction

Evolution and Natural Selection: When people hear the word "evolution," they often think of the Big Bang, dinosaurs, fossils, and events that occurred in the past. A scientist looks at evolution quite differently. Simply put, evolution is a change over time in the genetic composition of a population of organisms. Evolution unites all disciplines of the life sciences and is responsible for the diversity of organisms around us. The mechanism for evolution is natural selection—the process by which populations of organisms can change over time if individuals having certain, beneficial, and heritable traits leave more offspring than individuals lacking those traits.

Typically, natural selection is fostered by the addition of a selective pressure to an organism's environment. Selective pressures put the organisms present in that environment under some sort of stress. Those organisms that survive the selective pressure have traits that provided them with a survival advantage compared to the organisms that did not survive. The surviving organisms are able to pass down the traits to their offspring so that they might be successful as well.

We can easily appreciate evolution by observing organisms with short generation times or organisms that produce high numbers of progeny. A classic example of this is the evolution of the peppered moth, *Biston betularia*, in Industrial England (~1760–1830). Prior to the Industrial Revolution, most of the peppered moths had a light grayish coloration with black speckles. Those moths with lighter coloration were able to blend in with the light-colored lichen that resided on tree trunks as a form of camouflage. Eventually, the soot from the coal-burning factories of the Industrial Revolution coated the tree trunks and killed the lichen, thereby exposing the lighter-colored peppered moths to their predators—birds. Peppered moths that had inherited darker coloration patterns now had a selective advantage compared to the lighter-colored moths. They were camouflaged by the darker trees and, because they had the new survival advantage, they were able to pass the traits responsible for their darker coloration to their offspring until the darker peppered moths far outnumbered the lighter peppered moths.

The acquisition of antibiotic resistance by bacteria is another example of evolutionary phenomena. In this particular scenario, the introduction of antibiotics served as the selective pressure. Bacteria that had acquired traits to resist the antibiotics had a survival advantage compared to bacteria that did not acquire those traits. In some cases, antibiotic resistant strains of bacteria have appeared within 1 year (or less) from the introduction of the drug into the clinical setting. There are several antibiotics that can no longer be used in the clinical setting because so many bacteria have acquired resistance to them.

The common denominator behind the shift in the coloration of the peppered moth population due to the soot produced during the Industrial Revolution and resistance to antibiotics by bacteria is evolution. In many cases, resistance to selective pressures is due to the acquisition of a mutation(s). The soot, antibiotics, and mutagens that were initially described in Exercise 8 can all serve as selective pressures to generate mutations that enhance an organisms' survival. Once acquired, these mutations can then be propagated in a population of cells or organisms over time, enabling them to resist the selective pressure for generations.

Using Ultraviolet Light to Study Evolution: In today's laboratory, we are going to expose *S. marcescens* to a different type of selective pressure to study evolution—the mutagen, ultraviolet (UV) light. We will expose *S. marcescens* to wavelengths of UV light between 260 and 270 nm using a germicidal lamp. These types of lamps are commonly used to reduce the microbial populations in clinical, pharmaceutical, and laboratory settings when the facilities are not in use.

The purine and pyrimidine bases of DNA strongly absorb UV radiation in this range of wavelengths. As a review, thymine (T) and cytosine (C) are the pyrimidine bases, and adenine (A) and guanine (G) are the purine bases. The sequence of these bases is important for the triplet genetic code in the DNA that makes up the proteins of an organism. When two pyrimidines located next to each other on the same strand of DNA absorb UV light, they form an unusual cyclobutane (4-carbon) bond between their adjacent carbons. This bond links the two pyrimidines in the same strand of DNA tightly together to form a structure called a pyrimidine dimer. The dimer does not properly base pair with the complementary bases in the opposite strand of the DNA. When the DNA is copied, during DNA replication, the wrong bases get incorporated into the DNA across from the pyrimidine dimers, therefore creating the mutation(s). An accumulation of too many mutations in the DNA of bacteria is lethal to them.

Many microorganisms have repair systems that can eliminate the damage caused by UV light. One of these repair systems is the photoreactivation photolyase enzyme. This enzyme restores the dimerized bases of the pyrimidine dimers to their original shape by cutting the bonds that make up the cyclobutane rings. In the presence of longer wavelength light, such as visible light, the repair system becomes activated and reverses the lethal action of UV light.

As a marker for evolution, we are going to analyze the impact of UV light exposure on prodigiosin production. As you might recall from Exercise 7, the production of prodigiosin requires a complex branched biosynthetic pathway. Upon exposure to UV light, some mutations might be generated in the genes encoding the enzymes involved in this pathway. Alternatively, mutations might occur in genes responsible for controlling the biochemical pathway. Two results might occur if mutations are generated in these genes:

(1) Prodigiosin production ceases or decreases significantly, leading to a loss of red pigmentation (hypo-pigmentation).

(2) Prodigiosin production is enhanced significantly, leading to an increase in red pigmentation (hyper-pigmentation).

Future Directions for This Laboratory Course: Today, you will work with your laboratory group to generate UV light-induced hypo- and hyper-pigmented mutants of *S. marcescens*. Beginning next laboratory exercise and continuing over the next several exercises, you will work with your group to select and further characterize your hyper-pigmented mutant of *S. marcescens* (using a hypo-pigmented mutant that you select as a control—do you know what type of control?). You will be performing studies on your group's mutant using techniques that you have already mastered in the laboratory. Your goal is to attempt to enhance the production of prodigiosin by *S. marcescens* and gain some initial understanding of the environmental, genetic, and biochemical conditions required for this to occur.

The end result of the series of experiments that you will perform over the next several weeks will be to generate enough data to prepare your second formal laboratory report to discuss how you generated and initially characterized your hyper-pigmented mutant of *S. marcescens*.

Let's Put This to Practice!

You are going to work in groups of four to generate the hypo- and hyper-pigmented mutants of *S. marcescens* using the following protocol. Please turn on the UV light prior to beginning this protocol so that it may have appropriate time to warm up:

(1) Determine the optical density of the *S. marcescens* culture supplied to your group at 600 nm. Use the sterile nutrient broth + 1.0% maltose as your blank. The optimal density at 600 nm was ________.

(2) Prepare serial dilutions of the *S. marcescens* culture supplied to your group using the procedure described below. You need to do this in order to dilute the culture supplied to you so that you will be able to see and select individual colonies on the plates you are going to expose to UV light. Refer to the Figure 9.1 to guide you through this process:

(a) Using sterile technique, add 0.1 ml of the culture to 9.9 ml of 0.85% sterile saline.

(b) Close the lid of the diluted culture and gently shake it side to side (do not flip the tube upside down).

(c) Using sterile technique, remove 0.1 ml of the diluted culture you just made and add it to 9.9 ml of new 0.85% sterile saline.

(d) Close the lid of the diluted culture and gently shake it side to side (do not flip the tube upside down).

(e) Using sterile technique, remove 0.1 ml of the second diluted culture you just made and add it to 9.9 ml of new 0.85% sterile saline.

(f) Close the lid of the diluted culture and gently shake it side to side (do not flip the tube upside down). You will use this final diluted culture for the rest of the experiment.

Refer to the Figure 9.1. The cell suspension in the third tube represents a _____-fold dilution of the original culture. Why is this?

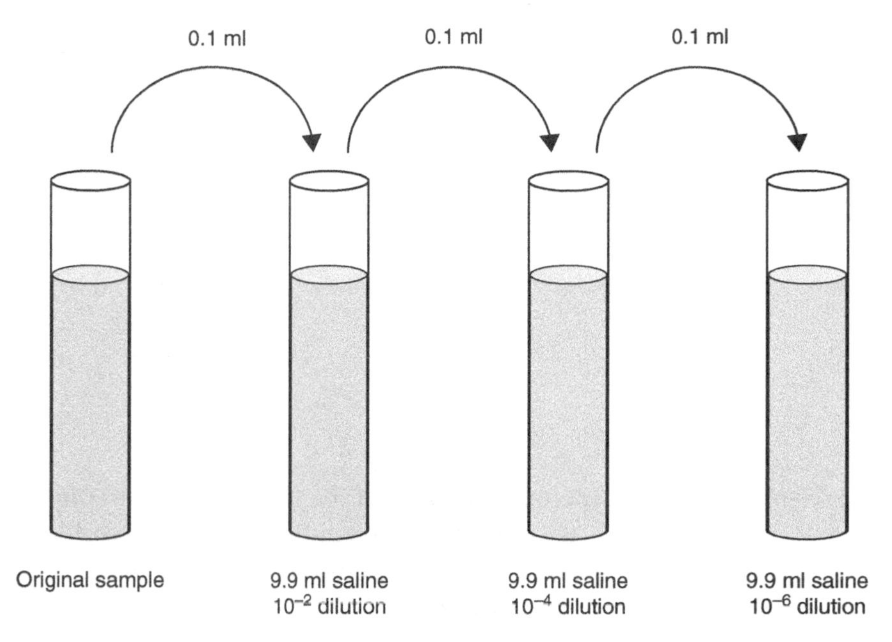

Figure 9.1 Serial dilutions. (*See insert for color representation of the figure*.)

(3) Determine the optical density of your diluted *S. marcescens* culture at 600 nm. Use the 0.85% sterile saline as your blank. The optimal density at 600 nm was _______ AU. Compare this number to the otpimal density you obtained for step 1 and your predicted fold dilution. Does the OD value you just measured make sense? You will need this information for your Post-laboratory Thinking Questions and Formal Laboratory Report 2.

(4) Label the perimeter of the bottom of six nutrient agar plus 1.0% maltose plates with your group name, your professor's name, the day of the week of your laboratory, and each of the following exposure times:

0 s (control)

10 s

20 s

30 s

1 min

1.5 min

(5) Using sterile technique, add 0.1 ml of the third dilution to one of the nutrient agar plus 1.0% maltose plates and use a cell spreader to make sure the entire medium surface is covered with bacteria. Your instructor will show you how to do this.

(6) Cover the plate and repeat step 3 using the other five plates.

(7) Place the six plates in the UV light box and remove the lids.

(8) Expose each plate for the prescribed time labeled on the plates. Once you remove the plates from the UV light box, immediately return the lid to the plate and place them in a closed drawer to limit exposure to visible light (Do you know why you need to do this?).

(9) Once you finish exposing all plates to UV light, wrap your plates in parafilm and place them back in the closed drawer. Make sure the drawer is labeled with your laboratory class professor's name and the day of week your laboratory runs.

(10) In the lab the following week, carefully examine the growth from your plates.

(11) Count the colonies on each plate using a colony counter and record the number of colonies per plate in Table 9.1.

(12) Count the number of red (normal), hyper-pigmented (bright red), and hypo-pigmented (white) colonies on each plate and record that information in Table 9.1, as well. To ascertain if a colony is considered hyper-pigmented, compare the suspected colonies to the red colonies present on the plate that was not exposed to UV light (exposure time = 0). If the suspected colonies are deeper red, consider the colony hyper-pigmented.

(13) As a group, select one hypo-pigmented colony from one of your plates and one hyper-pigmented colony from one of your plates by putting a circle around the colony on the back side of the plate using a marker. Have your instructor review your colony selections before you continue.

(14) Use sterile technique to inoculate 15 ml of fresh media with each colony (one culture per colony). Your instructor will show you how to do this.

(15) Label your group's two cultures with your group name, your professor's name, the day of the week your laboratory

Table 9.1 Pigment characteristics of *S. marcescens* colonies following UV light mutagenesis.

Exposure time	Total number of colonies	Number of normal red colonies	Number of hyper-pigmented colonies (bright red)	Number of hypo-pigmented colonies (white)
0 s				
10 s				
20 s				
30 s				
1 min				
1.5 min				

runs, and whether or not the culture contains the hypo- or hyper-pigmented mutant.

The laboratory prep staff will incubate your cultures on a shaking incubator set to 30°C so that you might continue to work with them during Exercise 11.

If you own your own laptop, please bring it to the lab next session. If you do not, please bring a device to save files.

NAME ______________________________ *DATE* _____________

Exercise 9C: Understanding Evolution by the Generation of UV Light-Induced Prodigiosin Mutants Post-laboratory Thinking Questions

Directions

Answer the following questions upon completion of the laboratory exercises.

For this laboratory, you used UV light to generate hypo- and hyper-pigmented mutants of *S. marcescens*. For your data analysis, you were first asked to count the number and type of colonies that grew on each plate and record that information in Table 9.1. Please ensure that you have all of the information in Table 9.1 to answer the following questions:

(1) In addition to the Table 9.1, you were asked to measure the OD_{600} of your *S. marcescens* culture before and after you diluted it (steps 1 and 3 in the protocol). Complete the middle column of Table 9.2 with those values.

(2) Haddix and colleagues (2000) determined that there are 5.8×10^8 bacteria/ml at an $OD_{600} = 1.0$ using a spectrophotometer set to 600 nm. Using this information and the optical densities you determined for the two cultures (Table 9.2), set up ratios to determine the number of bacteria/ml present in each of your cultures before and after you diluted them. Record these values in the last column of Table 9.2.

(3) On each plate, you plated 0.1 ml from your diluted *S. marcescens* culture. Each colony you counted on each plate following incubation is derived from a single bacterial cell (aka colony forming

Table 9.2 Number of bacteria/ml in initial and diluted *S. marcescens* cultures.

Culture type	Optical density at 600 nm (OD)	Number of bacteria/ml
Initial *S. marcescens* culture		
Diluted *S. marcescens* culture		

unit; CFU). You can use these two pieces of information and the number of bacteria/ml in your diluted culture (Table 9.2) to determine the percent survival upon exposure to UV light. Use the Table 9.3 to help you do this.

(4) What percentage of the bacteria that survived exposure to UV light under the different exposure times had mutations that affected prodigiosin production? To determine this, use the information in Tables 9.1 and 9.4 to assist you with the calculations.

(5) Review the data in Data Tables 9.3 and 9.4. What is the relationship between the percent of bacteria that survived exposure to UV light at the different exposure times and the number of bacteria that have mutations affecting prodigiosin production at the different exposure times? Support your conclusion with evidence from the data.

(6) (Optional) What is the mutation rate for the production of prodigiosin mutants (hyper- and hypo-pigmented mutants, together)? Review the materials in Exercise 8 and use the following procedure to assist you in determining the mutation rate:

> **(a)** For Table 9.2, you determined the number of bacteria/ml in your diluted *S. marcescens* culture. This concentration is concentration of bacteria that you exposed to UV light.

Table 9.3 Percent survival of *S. marcescens* after exposure to UV light.

Exposure time	Total # of counted colonies (copy from Table 9.1, second column)	Total # of bacteria/ml after exposure to UV light (multiply total # of counted colonies by 10)	Percent surviving {(total # of bacteria/ml after UV exposure/number of bacteria per ml present in your diluted culture—Data Table 9.2) × 100}
0 s			
10 s			
20 s			
30 s			
1 min			
1.5 min			

(b) In Table 9.3, you determined the total number of bacteria/ml present after UV exposure (third column) for each exposure time. Select the UV light exposure time with the highest percent survival (as per the results in the fourth column of Table 9.3) and determine the total number of doublings (events) that occurred. To do this, divide the total number of bacteria/ml present after UV exposure at the time point you selected (third column of Table 9.3) by the total number of bacteria/ml you exposed to UV light (Table 9.2). Record your answer here: (i) ____. This number represents the number of cell divisions (events) that occurred after exposure to the UV light.

(c) What was the number of mutants (hyper- and hypopigmented, combined) that you recovered at the exposure time with the highest percent survival (fourth column, Table 9.4)? Record that number here: (ii) _____. Multiply the number by 10 to determine the concentration (in bacteria/ml) of mutant bacteria recovered at the exposure time with the highest percent survival. Record that value here: (iii) _____. This number represents the total number of mutations resulting from each cell division event in that culture. Subtract this number from the total number of bacteria/ml present after

Table 9.4 Percentage of surviving *S. marcescens* with mutations affecting prodigiosin production.

Exposure time	Number of hyper-pigmented colonies (copy from Table 9.1, fourth column)	Number of hypo-pigmented colonies (copy from Table 9.1, fifth column)	Number of prodigiosin production mutants (# of hyper-pigmented colonies + # of hypo-pigmented colonies)	Total number of colonies (copy from Table 9.1, second column)	% of bacteria surviving UV light exposure that have mutations affecting prodigiosin production {(number of prodigiosin production mutants/total number of colonies) × 100}
0 s					
10 s					
20 s					
30 s					
1 min					
1.5 min					

UV exposure (Table 9.3, third column) for the exposure time with the highest percent survival. Record the result here: (iv) _____. This number represents that total number of bacteria that did not mutate in response to exposure to UV light.

(d) Use the steps below and the numbers you calculated above to determine the mutation rate for you cultures upon exposure to UV light:

Concentration of cells that did not mutate in bacteria/ml (iv): _____

N or total number of cell divisions (i): _____

$P_{12}(0)$ = concentration of cells that did not mutate/total number of cell divisions = _____

$\ln (P_{12}(0))$ = _____

$-\ln (P_{12}(0))/N$ = _____ (mutation rate)

(7) (Optional) How does your mutation rate after exposure to UV light compare to the spontaneous mutation rate for bacteria (one altered base per 10^{10} bases)? Does this make sense?

NAME ______________________________ *DATE* ____________

Exercise 10A: Understanding the Energy Spilling Properties of Prodigiosin Pre-laboratory Thinking Questions

Directions

Read over the introduction, protocols, and Haddix *et al.* (2008) for this laboratory exercise, and answer the following questions to ensure that you are prepared for the session.

Please note that the Haddix *et al.* (2008) paper might be difficult for you to follow. This is absolutely acceptable and OK. If you have questions, please see your instructor and he/she will be more than willing to assist you:

(1) What was the specific purpose of the study presented in Haddix *et al.* (2008)?

(2) What is the hypothesis for the study presented in Haddix *et al.* (2008)?

(3) Based upon the data obtained, what conclusions did the authors make?

(4) In the laboratory this week, you will be performing the ATP assay that is described in Haddix *et al.* (2008). You will be performing the assay using stationary phase wild-type, 933, and

The Fundamentals of Scientific Research: An Introductory Laboratory Manual, First Edition. Marcy A. Kelly.

Companion website: www.wiley.com\go\kelly\fundamentals

WCF *Serratia marcescens* cultures. Based upon the protocol presented in this laboratory manual, what are your hypotheses with respect to the amount of ATP produced by the three strains that you are testing?

NAME ______________________________ *DATE* ____________

Exercise 10B: Understanding the Energy Spilling Properties of Prodigiosin

Please remember to complete the work required for Exercise 9 prior to starting this laboratory.

Introduction

Cellular Respiration: Cellular respiration is an enzyme-mediated, catabolic process by which organisms convert the energy stored in the covalent bonds of organic molecules, such as glucose, into chemical energy in the form of ATP. The ATP that is produced by cellular respiration is used by the organisms' cells to perform the work of growth, maintenance, and function. The molecular formula for cellular respiration is as follows:

Equation 10.1

Molecular equation for cellular respiration:

$$C_6H_{12}O_6 + 6O_2 \rightarrow 6CO_2 + 6H_2O + 38ATP$$
$$\Delta G = -686 \text{ kcal/mol}$$

Although the molecular formula for cellular respiration seems simple, the process is actually broken down into three complex steps, each consisting of multiple enzyme-mediated reactions. The three steps are glycolysis, the Krebs or citric acid cycle, and the electron transport chain (ETC). Glycolysis is an anaerobic process that occurs in the cytoplasm of all organisms. Both the Krebs cycle and the ETC are aerobic processes. The Krebs cycle occurs in the cytoplasm of prokaryotic organisms and the mitochondrial matrix of eukaryotic organisms. The ETC occurs on the plasma membrane of prokaryotic organisms and on the mitochondrial inner membrane

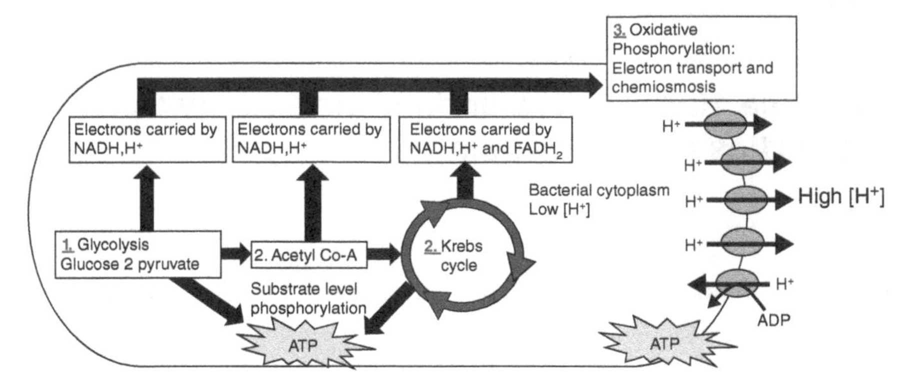

Figure 10.1 Bacterial cellular respiration. (*See insert for color representation of the figure.*)

in eukaryotic organisms. The three steps and their relationship in prokaryotic cells are summarized in Figure 10.1.

The first step, glycolysis, begins with a molecule of glucose. The glucose molecule is broken down into two pyruvate molecules through biochemical pathway with 10 enzyme-mediated reactions. These reactions also result in the formation of two additional molecules: ATP and NADH,H^+. Two ATP are generated through a mechanism called substrate-level phosphorylation in glycolysis. For substrate-level phosphorylation to occur, an enzyme must transfer a phosphate group directly from a substrate molecule to ADP. The two NADH,H^+ molecules are produced through oxidation–reduction reactions where electrons derived from glucose reduce two NAD^+ molecules. NAD^+ is an electron-shuttling molecule because when it is associated with electrons, they are shuttled to the site of the final step of cellular respiration—the ETC.

Before the Krebs cycle can begin, the two pyruvate molecules produced during glycolysis must be converted to acetyl-CoA. This step results in the reduction of another two NAD^+ molecules to form two more NADH,H^+ that travel to the site of the ETC. As the name implies, the Krebs cycle consists of a series of enzyme-mediated chemical reactions that are linked together in a continuous cycle of reactions. Three molecules of NADH,H^+, one molecule of $FADH_2$ (a second electron-shuttling molecule), and one molecule of ATP are produced from one turn of the cycle. The cycle must go around two times per glucose molecule. Therefore, in total, six molecules of NADH,H^+, two molecules of $FADH_2$, and two molecules of ATP are generated per one glucose molecule. The ATP generated by the Krebs cycle is produced by substrate-level phosphorylation. The NADH,H^+ and the $FADH_2$ produced both go to the site of the ETC.

During the ETC, the energy in the electrons carried by the NADH,H^+ and $FADH_2$ molecules is used to generate a proton gradient across either the mitochondrial inner membrane or the prokaryotic cell membrane. To do this, the energy from the electrons is used to pump hydrogen ions across the membranes, against their concentration gradients. Once the energy in the electrons is used up, the electrons associate with an oxygen molecule to form water (a by-product of cellular respiration). The oxygen molecule is considered the final electron acceptor for the ETC. This reliance on oxygen is why the ETC is considered an aerobic process. If the electrons donated to the ETC from NADH,H^+ and $FADH_2$ cannot associate with oxygen, the entire process gets backed up, and as a result, the Krebs cycle ceases to function as well.

The membranes on which the ETC occurs are impermeable to hydrogen ions. Therefore, as the hydrogen ions build up on one side of the membrane, they begin to store potential energy. The hydrogen ions can only get back across the membrane (down their concentration gradients) through a membrane-associated hydrogen ion transport complex called the ATP synthase. As the hydrogen ions pass through this complex, the energy that they yield is used to phosphorylate an ADP molecule. The process by which ATP is made through the transport of hydrogen ions down their concentration gradients is called chemiosmotic phosphorylation or chemiosmosis.

Table 10.1 Amount of ATP generated during each step of cellular respiration.

Step	Molecule produced	Number of ATP generated
Glycolysis	• 2 ATP by substrate-level phosphorylation • 2 NADH,H+	8 ATP
Conversion of pyruvate to acetyl-CoA	• 2 NADH,H^+	6 ATP
Krebs cycle	• 2 ATP by substrate-level phosphorylation • 6 NADH,H^+ • 2 $FADH_2$	24 ATP
Total ATP		*38 ATP*

Generally, for every NADH,H^+ molecule that enters the ETC, three ATP molecules are produced. For every $FADH_2$ molecule that enters the ETC, two molecules of ATP are produced. Therefore, as a result of cellular respiration, approximately 38 ATP are produced as per Table 10.1.

Prodigiosin Hypothesized to Have a Role in Energy Spilling. Haddix and colleagues (2008) demonstrated that prodigiosin synthesis has a negative regulatory role with respect to the production of ATP during the aerobic growth of *S. marcescens*. Based upon their work, they hypothesized that prodigiosin functions as an uncoupler of hydrogen ion transport and ATP synthesis. This implies that prodigiosin acts as a transporter on the bacterial cell membrane that transports hydrogen ions in the same direction as the ETC associated ATP synthase—it replaces the ATP synthase with respect to the transport of hydrogen ions. Simply put, it is proposed that prodigiosin dissipates the hydrogen ion gradient that was set up across the bacterial cell membrane by the ETC without producing ATP. This phenomenon is called energy spilling.

It is unclear as to why prodigiosin might have such an energy spilling function. You might recall that you observed that prodigiosin production increases as the cells progress through the bacterial growth curve in Exercise 5. Perhaps, once prodigiosin reaches a certain concentration, entry into stationary phase is triggered. One of the hallmarks of stationary phase is a decrease in ATP production—which may be due to the energy spilling properties of prodigiosin. What do you think? Can you come up with another hypothesis?

Luminescence Can Be Used to Evaluate ATP Production. In today's laboratory, you will work in groups to determine the amounts of ATP produced by stationary phase cultures of wild-type and the two prodigiosin null mutants of *S. marcescens* that you used in Exercise 6. To do this, you will use a kit from a biotech company. This kit includes a substrate molecule called luciferin (derived from a beetle) that binds to ATP molecules. The kit also includes a thermostable enzyme called luciferase. Luciferase is the enzyme produced by fireflies that enables them to emit light. In our case, the luciferase will be used to cleave luciferin when it is bound to ATP. Cleaved luciferin emits luminescent light which can be measured using a gel documentation system or a plate reader that detects luminescence. The luciferin/luciferase reaction is depicted in Figure 10.2. The amount of luminescent light emitted by each of your samples is proportional to the amount of ATP they produce.

Let's Put This to Practice!

Sterile technique is not required for this laboratory exercise, but you will need to wear gloves so that ATP from your hands does not contaminate the experiment.

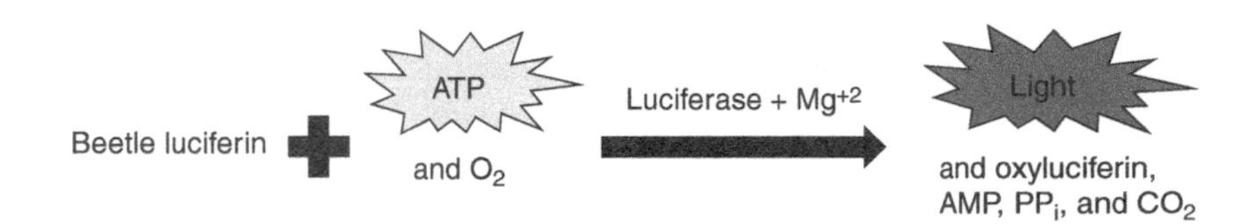

Figure 10.2 The luciferase reaction to quantify the amount of ATP produced by wild-type and mutant *S. marcescens*. (*See insert for color representation of the figure.*)

Each group will be working with the following samples, standards, and controls (Table 10.2; make sure you understand why each sample has the roles listed).

Table 10.2 Samples and controls for the ATP assay.

Sample	Role
Fresh, sterile nutrient broth + 1.0% maltose	Negative control, blank for standards
1 uM ATP	Standard
0.5 uM ATP	Standard
0.1 uM ATP	Standard
0.05 uM ATP	Standard
0.01 uM ATP	Standard
S. marcescens D1	Test
S. marcescens WCF	Positive control
S. marcescens 933	Positive control

The entire class will share one 96-well microplate for their samples, so each group will have to wait in turn to load their samples into the wells of the plate:

(1) Load 100 ul of each sample listed in Table 10.2 into the wells on the microplate. Make sure you write down which wells you inoculated and what sample is in each well.

(2) After each group has loaded their samples, each group should then load 100 ul of the luciferase reagent to each well that contains their sample.

(3) Incubate the microplate in a slow-shaking incubator at 23°C for 10 min.

(4) Your instructor will then take the plate and read/record luminescence using either the plate reader or gel documentation system.

Once you get the data, use a computer to create a standard curve using the blank and ATP standards. The units for the luminescent values you obtain from the plate reader or gel documentation system are relative luminescent units (RLU). If you need help with the generation of a standard curve using Excel, please refer to Exercise 3 in this laboratory manual. Set your *y*-intercept to 0 for your standard curve.

After you create the standard curve, determine the amounts of ATP produced in uM by *S. marcescens* D1, WCF, and 933 using Excel (how to determine x when you know y, Exercise 3).

If you own your own laptop, please bring it to lab next session. If you do not, please bring a device to save files.

NAME ______________________________ *DATE* ____________

Exercise 10C: Understanding the Energy Spilling Properties of Prodigiosin Post-laboratory Thinking Questions

Directions

Answer the following questions upon completion of the laboratory exercise:

(1) For this laboratory, you determined the amount of ATP produced by stationary phase *S. marcescens*. Why were *S. marcescens* WCF and 933 used as positive controls for this experiment?

(2) You were asked to generate a standard curve for this laboratory. Please copy/paste your standard curve into this worksheet.

(3) Complete Table 10.3 to indicate the concentrations of ATP produced by the three strains of *S. marcescens*.

(4) Do the results you obtained make sense? Why or why not? In your discussion, please remember to include your insights about the role of quorum sensing in prodigiosin production (quorum sensing is discussed in the introduction to Exercise 5) and information you learned from Haddix *et al.* (2008).

Table 10.3 Intracellular ATP concentration in wild-type and prodigiosin production mutants of *S. marcescens*.

Strain	ATP concentration (uM)
S. marcescens D1 (wild type)	
S. marcescens WCF	
S. marcescens 933	

MODULE 3

Initial Characterization of Novel *Serratia marcescens* Prodigiosin Mutants

NAME ______________________________ *DATE* ____________

Exercise 11A: Prodigiosin Mutant Study Part 1 Pre-laboratory Thinking Questions

Directions

Read over the introduction and information for this laboratory exercise, and answer the following questions to ensure that you are prepared for the session:

(1) What environmental conditions did you determine maximized the production of prodigiosin in Exercise 6 (temp, pH, aeration requirements, salinity, and carbon source)?

(2) Your goal for this laboratory is to determine if the environmental conditions you determined in Exercise 6 will further maximize the production of prodigiosin by your hyper-pigmented mutant. Please write up a detailed protocol for this week's laboratory using the environmental conditions you determined maximized prodigiosin production in Exercise 6 and the protocol for Exercise 5 to assist you. You will need to take readings over 3h (180min). Be sure to include controls in your protocol. Submit your protocol to your laboratory instructor for review and comment prior to your laboratory session as per his/her instructions.

(3) What are your hypotheses with respect to the growth and prodigiosin production by your hyper-pigmented mutant compared to wild-type and your hypo-pigmented mutant?

The Fundamentals of Scientific Research: An Introductory Laboratory Manual, First Edition. Marcy A. Kelly.

Companion website: www.wiley.com\go\kelly\fundamentals

EXERCISE 11B

NAME ______________________________ *DATE* ____________

Exercise 11B: Prodigiosin Mutant Study Part 1

Introduction

During Exercise 10, each group selected and inoculated one hypo-pigmented and one hyper-pigmented mutant of *S. marcescens* to characterize for the next 2 weeks. Although you will be using protocols that you have used in this laboratory in the past, the results that you will obtain are unknown to yourselves and your instructors. It is possible that all of your results might be negative (meaning that you end up with no growth, no prodigiosin production, and no reaction). Do not be discouraged if this happens to you! This is science! Most of the work we do in the laboratory produces negative data. It is in trying to figure out why the negative data was obtained that science really occurs.

Your goal over the next 2 weeks is to become scientific researchers. You will make and record observations about your mutants and then, based upon your observations, come up with a hypothesis as to why each of your mutants is behaving the way that they are. You will be asked to write up the data you obtain from Exercises 9, 11, and 12 in the format of a formal laboratory report. In your discussion, you will be responsible for stating the hypotheses you develop for each mutant and describing one possible experiment to test each of your hypotheses. In preparation for this, you might want to obtain and read the citations listed in the back of this laboratory manual. You also might want to perform some literature searches of your own, keeping in mind that good references only come from .edu, .org, and .gov websites (DO NOT USE WIKIPEDIA AS A PRIMARY SOURCE!).

Let's Put This to Practice!

Perform the growth and prodigiosin production experiment you developed for your pre-laboratory thinking questions. Record your results in Table 11.1. Perform the following prior to beginning the growth study:

(1) Obtain three 50 ml conical vials that contain 7.5 ml of fresh, sterile media. The conical vials will be labeled "growth study."

(2) Label each conical vial with the name of wild-type or your mutants.

(3) Using sterile technique, remove 7.5 ml from each of the three cultures (wild-type and the two mutant cultures you made last week) and put them into the appropriate "growth study" conical vial.

(4) Measure the absorbance/optical density of each of the newly made cultures at 499 and 600 nm.

(5) Record the measurements in the slot for time 0 in Table 11.1.

(6) Incubate the three cultures at 30°C with shaking for half an hour and follow your protocol to take readings. Record your readings in Table 11.1.

(7) During one of your incubation breaks, use the 100 ul from the two mutant cultures that you made last week (and that you used today for your studies) to inoculate 5 ml fresh, sterile nutrient broth + 1.0% maltose labeled "inoculate today." You will use these cultures for your studies next week. Be sure to label your cultures with the name of the organism, your group's name, your instructor's name, and the day of the week your lab runs.

Table 11.1 Absorbance at 499 nm and optical density at 600 nm for wild-type and prodigiosin production mutants of *S. marcescens* over 3 h (180 min).*

	Absorbance units at 499 nm (AU)			Optical density at 600 nm (OD)		
Time (min)	D1	HYPO	HYPER	D1	HYPO	HYPER
0						
30						
60						
90						
120						
150						
180						

* You might need to add additional information to this table based upon the environmental conditions you used.

Although not officially assigned until after next week's laboratory session, you might want to consider beginning your formal laboratory report now.

NAME ______________________________ *DATE* ____________

Exercise 11C: Prodigiosin Mutant Study Part 1 Post-laboratory Thinking Questions

Directions

Answer the following questions upon completion of the laboratory exercises.

For this laboratory, you began to characterize your hypo- and hyper-pigmented mutants of *S. marcescens*. You were first asked to perform a study similar to the one performed during Exercise 5 using the environmental condition that you identified in Exercise 6 as the condition resulting in optimal prodigiosin production. Please ensure that you have all of the information in Table 11.1 to answer the following questions:

(1) When you performed Exercise 5, you were asked to convert the absorbance units you obtained for the prodigiosin production study to units of prodigiosin production per bacterial cell. Please use those calculations to do the same and complete Table 11.2 (you might need to add more information to the table, depending upon the environmental conditions you used).

(2) Please create a growth curve multi-line graph using the data in Table 11.1 for the three strains you studied. Copy/paste your growth curve into this worksheet.

(3) Please create a unit prodigiosin produced per bacterial cell multi-line graph using the data in Table 11.2 for the three strains you studied. Copy/paste your graph into this worksheet.

Table 11.2 Amount of prodigiosin produced by wild-type and the prodigiosin production mutants of *S. marcescens* over 3 h (180 min).

	Prodigiosin (units/cell)		
Time (min)	D1	HYPO	HYPER
0			
30			
60			
90			
120			
150			
180			

(4) Based upon the aforementioned graphs and tables, what kinds of conclusions can you draw thus far with respect to the growth and prodigiosin production by your mutants compared to wild-type *S. marcescens*? Hint: When discussing the data in your graphs, use the slope values to indicate better growth/prodigiosin production (do you know why this is?).

(5) Do your conclusions match the hypotheses you proposed for the prelaboratory thinking questions? Why or why not?

NAME ______________________________ *DATE* ____________

Exercise 12A: Prodigiosin Mutant Study Part 2 Pre-laboratory Thinking Questions

Directions

Read over the introduction and information for this laboratory exercise and answer the following questions to ensure that you are prepared for the session:

(1) Please write up a detailed protocol for this week's laboratory using the protocol for Exercise 7 to assist you. Be surc to include controls in your protocol. Submit your protocol to your laboratory instructor for review and comment prior to your laboratory session.

(2) What are your hypotheses with respect to the impact of your two mutants on the feeding experiments?

The Fundamentals of Scientific Research: An Introductory Laboratory Manual,
First Edition. Marcy A. Kelly.

Companion website: www.wiley.com\go\kelly\fundamentals

NAME ______________________________ *DATE* ____________

Exercise 12B: Prodigiosin Mutant Study Part 2

Introduction

This week, you will continue with your mutant study by performing a feeding experiment similar to the one you worked on for Exercise 7. The goal is to determine if your hyper-pigmented mutant has an impact on the prodigiosin production by the wild-type and/or the null mutants and vice versa. You should realize that there are many explanations as to why your mutants are behaving the way they are. It might be that the mutants have defects in the enzymes of the biosynthetic pathway used to produce prodigiosin. It may also be that your mutants are defective in coloration due to mutations in genes that control the biosynthetic pathway such as the genes that encode the quorum sensing proteins. It might also be that the alterations in your mutants are in genes that are seemingly unrelated to prodigiosin production and that the coloration defects are secondary effects. Your laboratory instructor will work with each group individually to discuss your results and these possibilities, if appropriate.

Let's Put This to Practice!

Using sterile technique, the cultures provided to you in the laboratory, and the protocol you developed, inoculate the feeding plates by following the information in Table 12.1.

Next week, record your results in Table 12.2:

Table 12.1 Inoculations for feeding experiments with wild-type and prodigiosin production mutants of *S. marcescens*.

Plate number	Position 1	Position 2
1	D1	NA
2	WCF	NA
3	933	NA
4	HYPO	NA
5	HYPER	NA
6	HYPO	D1
7	HYPO	WCF
8	HYPO	933
9	HYPER	D1
10	HYPER	WCF
11	HYPER	933
12	HYPER	HYPO

Table 12.2 Results from feeding study with wild-type and prodigiosin production mutants of *S. marcescens*.

Position 1	Position 2	Color of bacterial growth
D1	NA	
WCF	NA	
933	NA	
HYPO	NA	
HYPER	NA	
HYPO	D1	
HYPO	WCF	
HYPO	933	
HYPER	D1	
HYPER	WCF	
HYPER	933	
HYPER	HYPO	

NAME ______________________________ *DATE* ____________

Exercise 12C: Formal Laboratory Report 2: Prodigiosin Mutant Study

This formal laboratory report will more closely mimic the reports that we prepare as professional scientists. You will be using the data you obtained from Labs 9, 11, and 12 for this report. Please review the introductions from Labs 11 and 12 for some more ideas about this report.

General Instructions

Formal laboratory reports must be typed (double spaced, 12-point Times New Roman font, default Word margins) and be between 20 and 25 pages. Tables and figures must be computer generated (not handwritten). You are expected to use proper grammar and correct spelling in your report. Therefore, please be sure to spell and grammar check your document before you submit it. Each section should begin on a new page and be written up as a clear, concise essay, not a list of answers to the points provided in this handout. The listed points should be used as a checklist to guide you through writing the lab report.

You will need to perform web-based research to learn more information than what is presented to you in this lab manual for your report (especially for the introduction section). When performing this research, only use sources from .edu, .org, or .gov websites. DO NOT use Wikipedia as a primary source. You will need to cite AT LEAST five sources (other than this laboratory manual) in your report. You may use some of the citations that are listed at the end of this lab manual.

Everything you write in your laboratory report should be in your own words. Summarize any web-based information, and then, cite the reference. Do not copy anything verbatim from any source (website, book, a peer, etc.)—using quotations in any form is not acceptable. For your citations, please follow the Council of Science Editors (CSE) citation style. You can most likely find information on this citation format in your college or university library.

Title: Laboratory reports should have a separate title page (include the title and your name). The title for your laboratory report should not be the title used in this laboratory manual; it needs to be more descriptive and include the following information in a single sentence (the information below is not listed in any particular order):

(a) The name of the organisms you worked with

(b) The phenotype that you were observing in each organism

(c) A concise statement summarizing a general conclusion you made from your data

Abstract: 250 words maximum:

(a) Name of the organisms you worked with.

(b) How you created the organisms.

(c) The conditions you tested.

(d) Briefly summarize your results.

EXERCISE 12C

Introduction: 3–5 pages. Web-based literature sources should be used to help you write this section:

(a) Background information about *Serratia marcescens*

(b) Background information about prodigiosin

(c) Background information on UV light mutagenesis and mutant selection

(d) Discussion of the detailed objectives/goals for your study

(e) Discussion of the experiments you performed to meet your objectives/goals

(f) Summary of your results

Materials and Methods: 1–2 pages. Provide details of what you did in the laboratory in paragraph format:

(a) Write at least 1 paragraph to describe each protocol you used for Exercises 9, 11, and 12. You might find that you will need more than one paragraph.

Results: Include the figures and narrative in the order as listed below. Use one figure per page in the report. Be sure to develop your own titles for each of your figures:

(a) Table 1—Number of colonies counted for each coloration (red, bright red, and white; Table 9.1).

(b) Table 2—Percent of bacteria surviving after exposure to UV light (Table 9.3).

(c) Table 3—Percent of bacteria surviving UV light exposure that also have mutations in prodigiosin production (Table 9.4).

(d) Figure 1—Graph of the growth of wild-type and the pigment mutants you selected.

(e) Figure 2—Graph of prodigiosin production by wild-type and the pigment mutants you selected.

(f) Table 4—Feeding experiment results for wild-type, 933, WCF, and the pigment mutants you selected (Table 12.2).

(g) Briefly summarize the information learned from all of the figures in paragraph format (1–2 pages). Complete this summary in three separate sections—focus the first section on the generation of the mutants (if you performed the optional analysis for Exercise 9, postlaboratory thinking questions, include your calculated mutation rate in your discussion), the second section on the optimal conditions required for growth, and the third section on the optimal conditions for prodigiosin production.

Discussion: 3–5 pages. You will have to do more web-based research and use more citations for this section:

(a) The first paragraph of your discussion should restate the objectives and hypotheses for your study and summarize how you performed the study.

(b) The next several paragraphs should describe your results again. You may use the summaries that you wrote in the results section of your report to assist you with this section, but shy away from directly copying and pasting from your results section. Summarize what you wrote in the results section.

(c) If you ran into any difficulties while you were performing your experiments that may have impacted your results, describe those difficulties here and suggest what impact you think they had on your data.

(d) Draw conclusions based upon each experiment you performed. When drawing your conclusions, refer to the data in the appropriate table/figure and include the actual numbers in the text of your discussion. Support each of your conclusions with data and/or citations from the literature. You should most definitely refer to what you wrote for the answers to the final question asked on the post-laboratory thinking questions for Exercises 9 and 11 to assist you.

(e) If you were to continue to work on this project, what would be your hypothesis for why your hypopigmented mutant is hypopigmented?

(f) Describe a single experiment you would like to perform to test this hypothesis. You will need to perform a literature search to help you with this. Make sure you cite what you find.

(g) If you were to continue to work on this project, what would be your hypothesis for why your hyper-pigmented mutant is hyperpigmented?

(h) Describe a single experiment you would like to perform to test this hypothesis. You will need to perform a literature search to help you with this. Make sure you cite what you find.

Citations: Use CSE format (see Appendix A). You should have at least five citations.

APPENDIX A

CSE Citation and Reference List Format Guidelines

The Council of Science Editors (CSE) publishes a general style guide for scientific writing. If you are planning on continuing your education in science or are planning on embarking on a career that involves a great deal of scientific writing, this is a good reference to have (a link to purchase this book from the CSE website is http://www.councilscienceeditors.org/publications/scientific-style-and-format/).

You are expected to use the CSE style to format all of your citations and reference lists used in your writing for this course. Furthermore, you should follow the name–year system. Links to information regarding how to format your references can be found at the following website:

http://www.lib.uoguelph.ca/get-assistance/writing/citations/cse-citation-name

The Fundamentals of Scientific Research: An Introductory Laboratory Manual,
First Edition. Marcy A. Kelly.

Companion website: www.wiley.com\go\kelly\fundamentals

APPENDIX B

Prodigiosin Biosynthesis

Excerpted from Bacterial Synergism Demonstration Kit (Item 154744), © Carolina Biological Supply Company. Used by permission only.

Serratia marcescens 933 produces a volatile precursor, 2-methyl-3-*n*-amyl-pyrrole (MAP), which is heavier than air. For this reason, the plates had to be incubated right side up, which allows MAP to collect on the agar instead of on the lid of an inverted Petri dish. MAP allows the *S. marcescens* WCF streaked opposite the 933 strain to complete the biochemical pathway and produce the red prodigiosin all along its length.

S. marcescens WCF produces a soluble precursor, 4-methoxy-2.2′-bipyrrole-5-carboxaldehyde (MBC), which diffuses through the agar and allows the *S. marcescens* 933 to complete the biochemical pathway and produce red prodigiosin in proximity to WCF.

S. marcescens D1 is used as the control because of its ability to complete the biochemical pathway and produce prodigiosin. The ability to produce prodigiosin is considered normal for *S. marcescens*, so D1 is genetically referred to as the wild-type strain.

A pyrrole is a ring compound that is the parent of many biologically important substances such as prodigiosin, bile pigments, porphyrins, chlorophyll, and hemoglobin. The structure of prodigiosin may be described by using the symbols A, B, and C to represent the three

The Fundamentals of Scientific Research: An Introductory Laboratory Manual, First Edition. Marcy A. Kelly.

Companion website: www.wiley.com\go\kelly\fundamentals

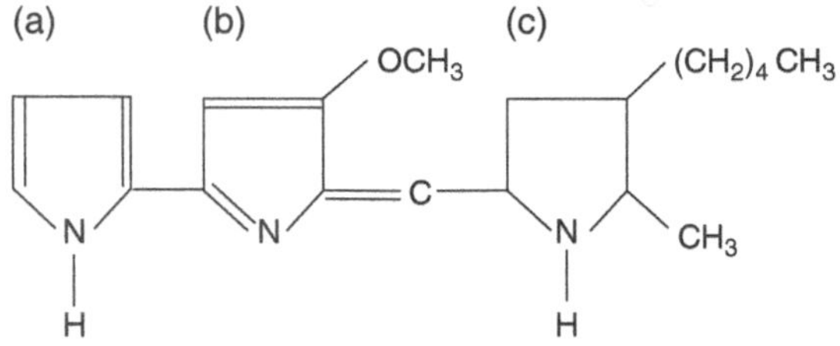

(a) (b) (c)

Figure B.1 Three pyrrole rings in the prodigiosin structure (labeled a, b, and c).

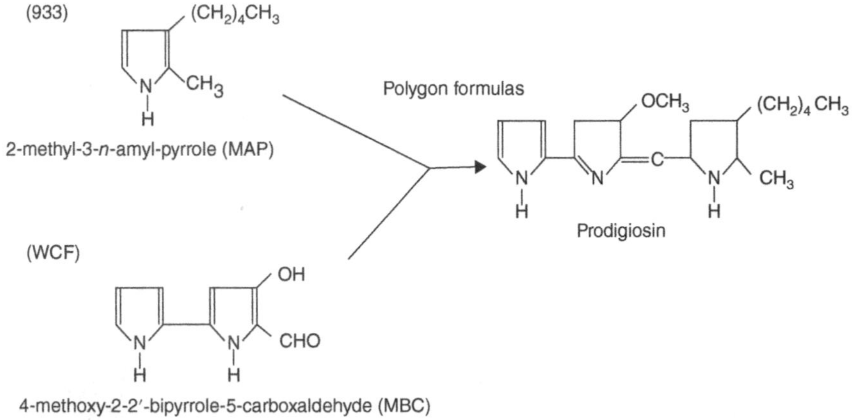

Figure B.2 Biochemical pathway responsible for the production of prodigiosin.

pyrrole structures in the molecule. All three pyrroles must be combined to yield the pigment prodigiosin (Figure B.1).

S. marcescens 933 produces and accumulates the volatile MAP (C in Figure B.1) but cannot complete the biosynthetic pathway because an enzyme is missing or nonfunctional. WCF produces and accumulates the MBC (A+B in Figure B.1) but also cannot complete the biochemical pathway because of a missing or nonfunctional enzyme. When 933 and WCF are streaked in proximity to each other on an agar dish, however, the two precursors, MAP and MBC, are then coupled, completing this biosynthetic pathway and forming the red pigment, prodigiosin (Figure B.2). The two mutants thereby supply each other with the missing precursors, allowing both to complete their biochemical pathways and produce prodigiosin.

References

American Association for the Advancement of Science. 2011. Vision and Change in Undergraduate Biology Education: A Call to Action. http://visionancchange.org/finalreport. Accessed March 31, 2015.

Bennett, JW. 1994. More on the Miracle of Bolsena. ASM News 60:403.

Cole, L. 2001. *The Eleventh Plague: The Politics of Chemical and Biological Warfare*. Diane Publishing Company, New York.

Cullen, JC. 1994. The Miracle of Bolsena. ASM News 60:187–191.

Haddix, PL., Jones, S., Patel, P., Burnham, S., Knights, K., Powell, JN., LaForm, A. 2008. Kinetic Analysis of Growth Rate, ATP and Pigmentation Suggests an Energy-Spilling Function for the Pigment Prodigiosin of *Serratia marcescens*. J. Bacteriol. 190(22):7453–7463.

Haddix, PL., Paulsen, ET., Werner, TF. 2000. Measurement of Mutation to Antibiotic Resistance: Ampicillin Resistance in *Serratia marcescens*. Bioscience 26(1):17–21.

Haddix, PL., Werner, TF. 2000. Spectrophotometric Assay of Gene Expression: *Serratia marcescens* Pigmentation. Bioscience 26(4):3–13.

Khanafari, A., Assadi, MM., Fakhr, FA. 2006. Review of Prodigiosin, Pigmentation in *Serratia marcescens*. Online J. Biol. Sci. 6(1):1–13.

Luria, SE., Delbrück, M. 1943. Mutations of Bacteria from Virus Sensitivity to Virus Resistance. Genetics 28(6):491–511.

The Fundamentals of Scientific Research: An Introductory Laboratory Manual, First Edition. Marcy A. Kelly.

Companion website: www.wiley.com\go\kelly\fundamentals

Index

The Fundamentals of Scientific Research: An Introductory Laboratory Manual,
First Edition. Marcy A. Kelly.

Companion website: www.wiley.com\go\kelly\fundamentals

INDEX